Machine Learning for Healthcare Systems: Foundations and Applications

RIVER PUBLISHERS SERIES IN COMPUTING AND INFORMATION SCIENCE AND TECHNOLOGY

Series Editors

K.C. CHEN
National Taiwan University,
Taipei, Taiwan

University of South Florida,
USA

SANDEEP SHUKLA
Virginia Tech,
USA

Indian Institute of Technology Kanpur,
India

The "River Publishers Series in Computing and Information Science and Technology" covers research which ushers the 21st Century into an Internet and multimedia era. Networking suggests transportation of such multimedia contents among nodes in communication and/or computer networks, to facilitate the ultimate Internet.

Theory, technologies, protocols and standards, applications/services, practice and implementation of wired/wireless

The "River Publishers Series in Computing and Information Science and Technology" covers research which ushers the 21st Century into an Internet and multimedia era. Networking suggests transportation of such multimedia contents among nodes in communication and/or computer networks, to facilitate the ultimate Internet.

Theory, technologies, protocols and standards, applications/services, practice and implementation of wired/wireless networking are all within the scope of this series. Based on network and communication science, we further extend the scope for 21st Century life through the knowledge in machine learning, embedded systems, cognitive science, pattern recognition, quantum/biological/molecular computation and information processing, user behaviors and interface, and applications across healthcare and society.

Books published in the series include research monographs, edited volumes, handbooks and textbooks. The books provide professionals, researchers, educators, and advanced students in the field with an invaluable insight into the latest research and developments.

Topics included in the series are as follows:-

- Artificial Intelligence
- Cognitive Science and Brian Science
- Communication/Computer Networking Technologies and Applications
- Computation and Information Processing
- Computer Architectures
- Computer Networks
- Computer Science
- Embedded Systems
- Evolutionary Computation
- Information Modelling
- Information Theory
- Machine Intelligence
- Neural Computing and Machine Learning
- Parallel and Distributed Systems
- Programming Languages
- Reconfigurable Computing
- Research Informatics
- Soft Computing Techniques
- Software Development
- Software Engineering
- Software Maintenance

For a list of other books in this series, visit www.riverpublishers.com

Machine Learning for Healthcare Systems: Foundations and Applications

Editors

C. Karthik
Jyothi Engineering College, Thrissur, India

M. Rajalakshmi
Sethu Institute of Technology, Madurai, India

Sachi Nandan Mohanty
VIT-AP University, Amaravati, AP, India

Subrata Chowdhury
Sreenivasa Institute of Technology and Management Studies,
Chittoor Andra Pradesh, India

Routledge
Taylor & Francis Group
NEW YORK AND LONDON

Published 2023 by River Publishers
River Publishers
Alsbjergvej 10, 9260 Gistrup, Denmark
www.riverpublishers.com

Distributed exclusively by Routledge
605 Third Avenue, New York, NY 10017, USA
4 Park Square, Milton Park, Abingdon, Oxon OX14 4RN

*Machine Learning for Healthcare Systems: Foundations and Applications /
C. Karthik, M. Rajalakshmi, Sachi Nandan Mohanty and Subrata Chowdhury.*

Routledge is an imprint of the Taylor & Francis Group, an informa business

ISBN 978-87-7022-811-4 (hardback)
ISBN 978-87-7022-905-0 (paperback)
ISBN 978-10-0095-998-7 (online)
ISBN 978-1-003-43881-6 (ebook master)

While every effort is made to provide dependable information, the publisher,
authors, and editors cannot be held responsible for any errors or omissions.

Contents

Preface   xi

List of Contributors   xiii

List of Figures   xvii

List of Tables   xxi

List of Abbreviations   xxv

1 Investigation on Improving the Performance of Class-imbalanced Medical Health Datasets   1

E. Ramanujam, T. Chandrakumar, and D. Sakthipriya

1.1 Introduction   2

1.2 Problem Formulation   3

1.3 Techniques to Handle Imbalanced Datasets   4

    1.3.1 Random undersampling (RUS)   4

    1.3.2 Random oversampling (RUS)   5

    1.3.3 Synthetic minority oversampling technique (SMOTE)   6

1.4 Classification Models   6

    1.4.1 Naive Bayes   6

    1.4.2 k-Nearest neighbor classifier   7

    1.4.3 Decision tree classifier   7

    1.4.4 Random forest   8

1.5 Dataset Collection   8

    1.5.1 Heart failure clinical records dataset   8

    1.5.2 Diabetes dataset   9

1.6 Experimental Results and Discussion   10

    1.6.1 Heart failure clinical records dataset   11

    1.6.2 Diabetes dataset   14

1.7 Conclusion   16

References   17

**2   Improving Heart Disease Diagnosis using Modified
    Dynamic Adaptive PSO (MDAPSO)**   **19**
    J. Shanthalakshmi Revathy, J. V. Anchitaalagammai, and
    S. Hariharasitaraman
    2.1   Introduction   20
    2.2   Background and Related Work   22
          2.2.1   PSO modification   22
          2.2.2   Feature selection.   23
          2.2.3   Classification and clustering.   23
    2.3   Proposed Approach   24
          2.3.1   PSO algorithm.   24
          2.3.2   Inertia weight   26
          2.3.3   Modified dynamic adaptive particle swarm
                  optimization (MDAPSO)   26
    2.4   Experimental Setup and Results.   27
          2.4.1   Dataset.   27
          2.4.2   Performance metrics.   28
          2.4.3   Result   29
    2.5   Conclusion.   32
    References.   32

3   Efficient Diagnosis and ICU Patient Monitoring Model   **35**
    Premanand Ghadekar, Pradnya Katariya, Shashank Prasad,
    Aishwarya Chandak, Aayush Agarwal, and Anupama Choughule
    3.1   Introduction   36
    3.2   Main Text   37
          3.2.1   Disease prediction   37
          3.2.2   Hospital monitoring system   41
    3.3   Experimentation.   47
          3.3.1   The threshold for the Levenshtein distance   47
          3.3.2   The threshold for heart rate and respiratory rate . . .   47
    3.4   Conclusion.   48
    References.   48

4   Application of Machine Learning in Chest X-ray Images   **51**
    V. Thamilarasi and R. Roselin
    4.1   Introduction   52
    4.2   Chest X-ray Images   54

4.3 Literature Review 55
4.4 Application of Machine Learning in Chest X-ray Images . . 57
    4.4.1 Clustering 58
    4.4.2 Regression 59
    4.4.3 Segmentation 61
    4.4.4 Classification 62
4.5 Case Study: Lung Chest X-ray Images 65
    4.5.1 Methodology 65
    4.5.2 JSRT dataset 66
    4.5.3 Image pre-processing 67
    4.5.4 CNN 68
4.6 Conclusion 69
4.7 Future Study 70
References 71

5 Integrated Solution for Chest X-ray Image Classification   **73**
Nam Anh Dao, Manh Hung Le, and Anh Ngoc Le
5.1 Introduction 73
5.2 Related Work 74
5.3 The Method 76
    5.3.1 Feature extraction 77
    5.3.2 Feature reduction 79
    5.3.3 Classification 81
    5.3.4 Algorithm 81
5.4 Experimental Results 82
5.5 Discussion and Conclusions 85
References 86

6 Predicting Genetic Mutations Among Cancer Patients by
Incorporating LSTM with Word Embedding Techniques   **91**
P. Shanmuga Sundari, J. Jabanjalin Hilda, S. Arunsaco and
J. Karthikeyan
6.1 Introduction 92
6.2 Related Work 93
    6.2.1 Basic feature engineering 93
    6.2.2 Classification method 95
6.3 Data Overview 95
    6.3.1 Dataset structure 95
    6.3.2 Data pre-processing 95

6.3.3 Most frequent genes and class. 95
6.3.4 Most frequent variation and class 96
6.3.5 Text length distribution and class 97
6.3.6 Word cloud to visualize data 97
6.4 Methodology 97
6.4.1 Pre-processing and feature extraction 97
6.4.2 Word embedding 98
6.4.3 Classifier. 101
6.5 Experiments 102
6.6 Results. 103
6.7 Evaluation Parameters. 104
6.8 Conclusion and Future Work 105
References. 106

7 **Prediction of Covid-19 Disease using Machine-learning-based Models** **109**

*Abhisek Omkar Prasad, Megha Singh, Pradumn Kumar Mishra,
Shubhi Srivastava, Dibyani Banerjee, and
Abhaya Kumar Sahoo*

7.1 Introduction 110
7.2 Literature Survey 111
7.3 Different Models used in Covid-19 Disease Prediction. . . . 112
7.3.1 Holt's linear model 112
7.3.2 Holt–Winters method 112
7.3.3 Linear regression 113
7.3.4 Polynomial regression. 114
7.3.5 Support vector machine 114
7.3.6 Moving average model 115
7.3.7 Autoregressive model 115
7.3.8 ARIMA 115
7.4 Evaluation Parameters used in Models 116
7.4.1 Root mean square error 116
7.4.2 Mean square error 116
7.4.3 Mean absolute error 117
7.5 Experimental Result Analysis 117
7.5.1 Future forecasting of death rates 118
7.5.2 Future forecasting of confirmed cases. 118
7.5.3 Future forecasting of recovery rate 121
7.6 Conclusion. 123
References. 127

8 Intelligent Retrieval Algorithm using Electronic Health Records for Healthcare   **131**

S. Nagarjuna Chary, N. Vinoth, and Kiran Chakravarthula

8.1   Introduction 132
8.2   EHR Datasets 133
     8.2.1   EHR repositories 134
8.3   Machine Learning Algorithm 135
     8.3.1   Data mining techniques and algorithms 136
     8.3.2   Machine learning algorithms using EHR for cardiac disease prediction 136
8.4   Machine Learning and Wearable Devices 138
8.5   Studies based on Data Fusion 139
8.6   Data Pre-processing 140
8.7   Conclusion 140
References 141

9 Machine Learning-based Integrated Approach for Cancer Microarray Data Analysis   **149**

Amrutanshu Panigrahi, Manoranjan Dash, Bibhuprasad Sahu, Abhilash Pati, and Sachi Nandan Mohanty

9.1   Introduction 150
9.2   Related Work 152
9.3   Background Study 154
     9.3.1   Machine learning 154
     9.3.2   Microarray data 158
9.4   Proposed Work 159
     9.4.1   RFE 159
     9.4.2   Cuckoo search 160
     9.4.3   Dataset 162
9.5   Empirical Analysis 162
9.6   Conclusion 163
References 165

10 Feature Selection/Dimensionality Reduction   **169**

Divya Stephen, S. U. Aswathy, Priya P. Sajan, and Jyothi Thomas

10.1   Introduction 170
10.2   Feature Selection 171
     10.2.1   Characteristics 172
     10.2.2   Classification of feature selection methods 173
     10.2.3   Importance of feature selection in machine learning . 177

    10.3 Dimensionality Reduction.   178
        10.3.1 Techniques for dimensionality reduction   179
        10.3.2 Advantages of dimensionality reduction   182
        10.3.3 Disadvantages of dimensionality reduction   182
    10.4 Conclusion.   182
    References   183

**11 Information Retrieval using Set-based Model Methods,
   Tools, and Applications in Medical Data Analysis**   **187**
   S. Castro, P. Meena Kumari, S. Muthumari, and J. Suganthi
    11.1 Introduction   188
    11.2 Literature Review   189
    11.3 Set-based Model for Reinforcement Learning Design for
        Medical Data   193
        11.3.1 Case study   195
        11.3.2 Rank computation   198
        11.3.3 Tools for evaluation   199
        11.3.4 Applications   200
    11.4 Conclusion.   201
    References   201

Index   **203**

Author Biographies   **205**

About the Editors   **221**

Preface

Healthcare is a significant sector that provides value-based services to millions of people while also generating significant income for several countries. Efficiency, importance, and result are three catchphrases that often surround healthcare and guarantee a lot; and today, healthcare experts and stakeholders all over the world are searching for new ways to achieve these goals. Smart healthcare allowed by technology is really no longer a pipe dream, as Internet-connected medical equipment are keeping the healthcare sector as we know it from collapsing under the weight of the community. Machine learning in medical care is one such field that is gradually gaining traction in the industry. Researchers at Stanford University are now using deep learning to detect lung cancer, and Google recently created a machine learning approach to determine cancerous tumors in colonoscopies. Machine learning is now helping in a variety of healthcare conditions. In healthcare, machine learning aids in the analysis of thousands of different data points and the prediction of outcomes. The importance of machine learning in medical services is its ability to process huge databases above human intellect and then efficiently turn statistical analysis into clinical insights that assist physicians in preparing and delivering care, eventually leading to improved results, lower healthcare costs, and improved customer satisfaction. Machine learning in medicine has recently gotten a lot of attention. Some systems lend themselves better to machine learning than others. Algorithms may help disciplines that have repeatable or structured processes right away. Those in fields like radiology, cardiology, and pathology that have massive image datasets are also good candidates. This book is a one-of-a-kind collection of strategies for representing, enhancing, and empowering interdisciplinary and multi-institutional machine learning research in health informatics. This book is a one-of-a-kind collection of current and evolving machine learning concepts for healthcare informatics, due to the diverse, uncertainty, and depth of this field. This book takes you on a journey of machine learning algorithms, architecture design, and medical applications. The future of machine learning in population and patient medical optimization will be explored, as well as the ethical consequences of machine learning in healthcare.

Editor

List of Contributors

Agarwal, Aayush, *Vishwakarma Institute of Technology, India*

Anchitaalagammai, J. V., *Department of Computer Science and Engineering, Velammal College of Engineering and Technology, India*

Arunsaco, S., *Department Mechanical Engineering, SVCET (A), India*

Aswathy, S. U., *Department of Computer Science and Engineering, Jyothi Engineering College, India*

Banerjee, Dibyani, *School of Computer Engineering, KIIT Deemed to be University, India*

Castro, S., *Department of Information Technology, Karpagam College of Engineering, India*

Chakravarthula, Kiran, *Department of Electronics & Instrumentation Engineering, VNR Vignana Jyothi Institute of Engineering and Technology, India*

Chandak, Aishwarya, *Vishwakarma Institute of Technology, India*

Chandrakumar, T., *Department of Applied Mathematics and Computational Science, Thiagarajar College of Engineering, India*

Chary, S. Nagarjuna, *Department of Electronics & Instrumentation Engineering, VNR Vignana Jyothi Institute of Engineering and Technology, India*

Choughule, Anupama, *Vishwakarma Institute of Technology, India*

Dao, Nam Anh, *Electric Power University, Vietnam*

Dash, Manoranjan, *Department of Management Sciences, Siksha 'O' Anusandhan (Deemed to be University), India*

Ghadekar, Premanand, *Vishwakarma Institute of Technology, India*

Hariharasitaraman, S., *School of Computing Science and Engineering, VIT Bhopal University, India*

Hilda, J. Jabanjalin, *Jabanjalin Hilda, School of Computer Science and Engineering, VIT University, India*

Hung Le, Manh, *Electric Power University, Vietnam*

Karthikeyan, J., *Department of Humanities and Science, SVCET(A), India*

Katariya, Pradnya, *Vishwakarma Institute of Technology, India*

Le, Anh Ngoc, *Swinburne Vietnam, FPT University, Vietnam*

Meena Kumari, P., *Department of Computer Science and Engineering, AVN Institute of Engineering and Technology, India*

Mishra, Pradumn Kumar, *School of Computer Engineering, KIIT Deemed to be University, India*

Mohanty, Sachi Nandan, *VIT-AP University, Amaravati, AP, India*

Muthumari, S., *Department of Computer Science & Information Technology, S.S. Duraisamy Nadar Mariammal College, India*

Panigrahi, Amrutanshu, *Department of CSE, Siksha 'O' Anusandhan (Deemed to be University), India*

Pati, Abhilash, *Department of CSE, Siksha 'O' Anusandhan (Deemed to be University), India*

Prasad, Abhisek Omkar, *School of Computer Engineering, KIIT Deemed to be University, India*

Prasad, Shashank, *Vishwakarma Institute of Technology, India*

Ramanujam, E., *Department of Computer Science and Engineering, National Institute of Technology Silchar, India*

Revathy, J. Shanthalakshmi, *Department of Computer Science and Engineering, Velammal College of Engineering and Technology, India*

Roselin, R., *Department of Computer Science, Sri Sarada College for Women (Autonomous), India*

Sahoo, Abhaya Kumar, *School of Computer Engineering, KIIT Deemed to be University, India*

Sahu, Bibhuprasad, *Department of Computer Science and Engineering, Gandhi Institute for Technology, India*

Sajan, Priya P., *C-DAC, India*

Sakthipriya, D., *Department of Applied Mathematics and Computational Science, Thiagarajar College of Engineering, India*

Singh, Megha, *School of Computer Engineering, KIIT Deemed to be University, India*

Srivastava, Shubhi, *School of Computer Engineering, KIIT Deemed to be University, India*

Stephen, Divya, *Department of Computer Science and Engineering, Jyothi Engineering College, India*

Suganthi, J., *Department of Information Science and Engineering, T. John Institute of Technology, India*

Sundari, P. Shanmuga, *Department of Computer Science and Engineering (Data Science), SVCET (A), India*

Thamilarasi, V., *Department of Computer Science, Sri Sarada College for Women (Autonomous), India*

Thomas, Jyothi, *Department of Computer Science and Engineering, Christ University, India*

Vinoth, N., *Department of Instrumentation, MIT Campus, Anna University, India*

List of Figures

Figure 1.1   Random undersampling technique   4

Figure 1.2   Random oversampling technique..   5

Figure 1.3   Bayes conditional probability equation.   7

Figure 1.4   Metadata information of heart failure clinical records dataset.   9

Figure 1.5   Distribution frequency of binary features – heart failure clinical records dataset.   9

Figure 1.6   Density values of discrete features – heart failure clinical records dataset.   10

Figure 1.7   Metadata information of diabetes dataset.   10

Figure 1.8   Density values of discrete features – diabetes dataset.   11

Figure 1.9   Distribution frequency of binary features – heart failure clinical records dataset.   11

Figure 2.1   Detailed flow of operation.   25

Figure 2.2   Accuracy in Cleveland dataset.   29

Figure 3.1   User interface to enter the symptoms.   38

Figure 3.2   JSON format with listed parameters (thyroid).   38

Figure 3.3   Uploading medical test reports with the required type.   40

Figure 3.4   Sample test format of PDF type (Bilirubin test). . . .   40

Figure 3.5   UI showing predicted disease, criticality level, and detailed analysis of the report.   41

Figure 3.6   Patient monitor display while extracting vital signs. .   42

Figure 3.7   The graph represents the actual heart rate values of the patient and predicted values by the model.   43

Figure 3.8   The graph represents the actual heart rate values of the patient and predicted values by the model.   44

Figure 3.9   Systolic blood pressure and age for male and female.   44

Figure 3.10   Diastolic blood pressure and age for male and female.   45

Figure 3.11 Patient profile page with an option to change the threshold value for a particular disease. 46

Figure 3.12 Alert message sent to hospital staff. 47

Figure 4.1 Simple structure of neural network.. 53

Figure 4.2 Types of chest X-ray images.. 55

Figure 4.3 Machine learning techniques.. 58

Figure 4.4 *k*-Means clustering.. 59

Figure 4.5 Linear regression.. 60

Figure 4.6 Model of decision tree. 60

Figure 4.7 Model of SVM. 61

Figure 4.8 Model of *k*-NN. 61

Figure 4.9 Nodule and non-nodule classification of SVM.. 63

Figure 4.10 Nodule and non-nodule classification by naïve Bayes.. 63

Figure 4.11 Accuracy from decision tree classifiers. 64

Figure 4.12 Methodology of machine learning. 66

Figure 4.13 Pre-processing in a lung CXR image. 68

Figure 4.14 Semantic segmentation of lung CXR image. 70

Figure 5.1 (1) Relation between chest X-ray image and class. (2). Image feature is determined by CNN on and then is reduced to have 78

Figure 5.2 Proposed schema of classification for chest X-ray images. 82

Figure 5.3 Algorithm of classification for chest X-ray images. 83

Figure 5.4 Distribution of classes in the chest X-ray image database. 84

Figure 5.5 Distribution of classes in the chest X-ray image database. 85

Figure 6.1 Most frequent genes and class. 96

Figure 6.2 Text length distribution and class. 98

Figure 6.3 Word cloud displaying most common terms.. 99

Figure 6.4 Proposed architecture. 100

Figure 6.5 LSTM architecture.. 102

Figure 6.6 Principle component analysis graph. 103

Figure 6.7 Performance of the classifiers. 104

Figure 7.1 Graph that represents line of regression. 113

Figure 7.2 Death cases forecasting graph using linear regression.. 118

Figure 7.3   Death cases prediction graph using polynomial
regression..   119
Figure 7.4   Death cases prediction graph using SVM.   119
Figure 7.5   Death cases prediction graph using the Holt's
linear model.   119
Figure 7.6   Death cases forecasting graph using the
Holt–Winters model.   120
Figure 7.7   Death cases prediction graph using the
AR model..   120
Figure 7.8   Death cases prediction graph using
the MA model.   120
Figure 7.9   Death cases prediction graph using the
ARIMA model.   121
Figure 7.10   Death cases prediction graph using the
SARIMA model.   121
Figure 7.11   Confirmed cases prediction graph using linear
regression..   122
Figure 7.12   Confirmed cases forecasting graph using
polynomial regression.   122
Figure 7.13   Confirmed cases prediction graph using SVM.. . . .   122
Figure 7.14   Confirmed cases prediction graph using the
Holt's linear model..   123
Figure 7.15   Confirmed cases prediction graph using the
Holt–Winters model.   123
Figure 7.16   Confirmed cases prediction graph using the
AR model..   123
Figure 7.17   Confirmed cases prediction graph using
the MA model.   124
Figure 7.18   Confirmed cases forecasting graph using the
ARIMA model..   124
Figure 7.19   Confirmed cases forecasting graph using the
SARIMA model.   124
Figure 7.20   Recovered cases prediction graph using linear
regression..   125
Figure 7.21   Recovered cases prediction graph using
polynomial regression.   125
Figure 7.22   Recovered cases prediction graph using SVM.. . . .   125
Figure 7.23   Recovered cases prediction graph using the
Holt's linear model..   125

Figure 7.24	Recovered cases prediction graph using the Holt–Winters model.	126
Figure 7.25	Recovered cases prediction graph using the AR model..	126
Figure 7.26	Recovered cases prediction graph using the MA model.	126
Figure 7.27	Recovered cases prediction graph using the ARIMA model.	126
Figure 7.28	Recovered cases prediction graph using the SARIMA model.	127
Figure 8.1	Year-wise publication in deep EHR. Data from Google scholar till September 2020.	133
Figure 8.2	EHR process hierarchy.	134
Figure 8.3	Classification of machine learning algorithms.	135
Figure 8.4	(A) Transmission and reflectance-mode photoplethysmography (PPG). (B) Variation in light attenuation by tissue.	139
Figure 9.1	Workflow of the proposed work.	161
Figure 9.2	LOO validation comparison among the proposed work, SVM-RFE, and SVM.	163
Figure 9.3	20-fold validation comparison among the proposed work, SVM-RFE, and SVM.	164
Figure 9.4	10-fold validation comparison among the proposed work, SVM-RFE, and SVM.	164
Figure 9.5	Fivefold validation comparison among the proposed work, SVM-RFE, and SVM.	165
Figure 10.1	Major stages in the feature selection process.	172
Figure 10.2	Feature selection illustration.	172
Figure 10.3	Taxonomy for feature selection.	174
Figure 10.4	Filter-based feature selection..	174
Figure 10.5	Wrapper-based feature selection.	175
Figure 10.6	Hybrid (filter- and wrapper-based feature selection).	175
Figure 10.7	Embedded or intrinsic-based feature selection.. . . .	176
Figure 10.8	Overview of filter, wrapper, and embedded methods. .	178
Figure 10.9	LDA, autoencoder, and *t*-SNE.	181
Figure 10.10	PCA vs. LDA..	181
Figure 11.1	Document representation	196
Figure 11.2	Modified term representation..	196
Figure 11.3	Text embedding for the selected textual data..	200

List of Tables

Table 1.1   Ten-fold cross-validation on imbalanced heart failure clinical records. 12

Table 1.2   Training and testing on 70-30 split of imbalanced heart failure clinical records. 12

Table 1.3   Ten-fold cross-validation on imbalanced heart failure clinical records using random undersampling. 13

Table 1.4   Training and testing on 70-30 split of imbalanced heart failure clinical records using random undersampling. 13

Table 1.5   Ten-fold cross-validation on imbalanced heart failure clinical records using random oversampling. 13

Table 1.6   Training and testing on 70-30 split of imbalanced heart failure clinical records using random oversampling. 13

Table 1.7   Ten-fold cross-validation on imbalanced heart failure clinical records using SMOTE. 14

Table 1.8   Training and testing on 70-30 split of imbalanced heart failure clinical records using SMOTE. 14

Table 1.9   Ten-fold cross-validation on imbalanced diabetes dataset. 15

Table 1.10   Training and testing on 70-30 split of imbalanced diabetes dataset. 15

Table 1.11   Ten-fold cross-validation on imbalanced diabetes dataset using random undersampling. 15

Table 1.12   Training and testing on 70-30 split of imbalanced diabetes dataset using random undersampling. 15

Table 1.13   Ten-fold cross-validation on imbalanced diabetes dataset using random oversampling. 16

Table 1.14   Training and testing on 70-30 split of imbalanced diabetes dataset using random oversampling. 16

Table 1.15   Ten-fold cross-validation on imbalanced diabetes
dataset using SMOTE.   16

Table 1.16   Training and testing on 70-30 split of imbalanced
diabetes dataset using SMOTE.   16

Table 2.1   List of features after feature extraction.   28

Table 2.2   Performance metrics.   29

Table 2.3   Cleveland dataset.   30

Table 2.4   Hungarian dataset.   30

Table 2.5   Switzerland dataset.   31

Table 2.6   VA dataset.   31

Table 3.1   Standard vital sign range.   42

Table 3.2   BP range for blood-pressure-related disease.   45

Table 3.3   SpO$_2$ variation considering lung disease.   46

Table 3.4   Calculating the Levenshtein distance for five
different reports and ratios.   47

Table 3.5   Comparing the errors of random forest regression
and decision tree.   48

Table 3.6   Comparing the errors of random forest regression
and decision tree for respiratory rate.   48

Table 4.1   Accuracy of decision tree classifiers.   64

Table 5.1   Dataset for the experiments.   83

Table 5.2   Performance report by accuracy.   84

Table 5.3   Performance report of related work.   86

Table 6.1   Evaluation metrics results.   103

Table 7.1   Models performance for predicting death rates
in the future.   118

Table 7.2   Models performance for predicting confirmed
cases in the future.   121

Table 7.3   Model performance for predicting the recovery
rate in the future.   124

Table 8.1   Data mining techniques and algorithms.   136

Table 8.2   Important attributes for heart disease prediction. . . .   136

Table 8.3   List of research papers on cardiovascular
health study.   137

Table 9.1   Related works.   155

Table 9.2   Dimension of the used dataset.   162

Table 9.3   Cross-validation for the proposed work with
different cancer datasets.   162

Table 9.4 Leave-one-out validation for the proposed work
with different cancer datasets. 163

Table 11.1 Document terms representation. 195

Table 11.2 Termsets and their corresponding document
classification. 197

Table 11.3 Weight computation and its termset representation. . 199

List of Abbreviations

AI	Artificial intelligence
ANN	Artificial neural network
AP	Anterior–posterior
API	Application programming interface
AR	Autoregressive
ARIMA	Autoregressive combined moving average
AUC	Area under curve
BAC	Bacterial artificial chromosome
BP	Blood pressure
BPSO	Binary particle swarm optimization
CAD	Computer-aided diagnosis
CAD	Coronary artery disease
CBOW	Continuous bag-of-words
CFA	Confirmatory factor analysis
CFS	Correlation-based feature selection
CHD	Coronary heart disease
CNN	Convolution neural network
COPD	Chronic obstructive pulmonary disease
CS	Cuckoo search
CT	Computed tomography
CXR	Chest X-ray
DICOM	Digital imaging and communications in medicine
DL	Deep learning
DT	Decision tree
DWT	Discrete wavelet transform
ECG	Electrocardiogram
EFA	Exploratory factor analysis
EHR	Electronic health record
EMR	Electronic medical record
EOG	Electrooculogram
ERP	Enterprise resource planning
FCM	Fuzzy C-means

FN	False negative
FP	False positive
GA	Genetic algorithm
GAN	Generative adversarial network
GEC	Grammatical error correction
GIST	Generalized search tree
GLM	Generalized linear model
GPR	Ground penetrating radar
GPU	Graphic processing unit
HITECH	Health Information Technology for Economic and Clinical Health
HR	Heart rate
ICA	Independent component analysis
IDF	Inverse document frequency
IDF	Inverse term frequency
IoT	Internet of Things
IR	Information retrieval
JSRT	Japanese Society of Radiological Technology
k-**NN**	*k*-Nearest neighbor
LDA	Latent Dirichlet allocation
LOO	Leave one out
LR	Linear regression
LSTM	Long short-term memory
MA	Moving average
MAE	Mean absolute error
MAP	Maximum *a posteriori*
MAPK	Mitogen activated protein kinases
MDAPSO	Modified dynamic adaptive particle swarm optimization
MEMS	Micro-electromechanical Systems
ML	Machine learning
MLPNN	Multilayer perceptron neural network
MRI	Magnetic resonance imaging
mRMR	Minimum redundancy maximum relevance
MSE	Mean squared error
MSKCC	Memorial Sloan-Kettering Cancer Center
MT	Microarray Technology
NB	Naïve Bayes
NCI	National Cancer Institute
NICE	National Institute for Health and Care Excellence
NIH	National Institutes of Health

NLP	Natural language processing
OCR	Optical character recognition
PA	Posterior–anterior
PCA	Principal component analysis
PDF	Portable Document Format
PNG	Portable network graphics
PNN	Polynomial neural network
PPG	Photoplethysmogram
PR	Polynomial regression
PSO	Particle swarm optimization
ReLU	Rectified linear unit
RF	Random forest
RFE	Recursive feature elimination
RL	Reinforcement learning
RMSE	Root mean square error
RNN	Recurrent neural network
ROC	Receiver operating characteristic
RR	Respiratory rate
RT-PCR	Real-time polymerase chain reaction
RUS	Random oversampling
RUS	Random undersampling
SARIMA	Seasonal autoregressive integrated moving average
SARS	Severe acute respiratory syndrome
SGD	Stochastic gradient descents
SMOTE	Synthetic minority oversampling technique
SNP	Single nucleotide polymorphism
SPLSDA	Sparse Partial Least Square discriminant analysis
SVM	Support vector machine
TB	Tuberculosis
TF	Term frequency
TN	True negative
TP	True positive
UCI	University of California Irvine
UNET	U-shaped encoder-decoder network architecture
VGG	Visual geometry group
VIF	Variance inflation factor
VSM	Vector space model
WDM	Wavelength Division Multiplexing
WHO	World Health Organization

1

Investigation on Improving the Performance of Class-imbalanced Medical Health Datasets

E. Ramanujam[1], T. Chandrakumar[2], and D. Sakthipriya[3]

[1]Department of Computer Science and Engineering, National Institute of Technology Silchar, India.
[2]Department of Applied Mathematics and Computational Science, Thiagarajar College of Engineering, India.
[3]Department of Applied Mathematics and Computational Science, Thiagarajar College of Engineering, India.
Email: ramanujam@cse.nits.a.cin; tckcse@tce.edu; sakthimca2011@gmail.com

Abstract

Data has increased significantly in recent years due to technical and technological advancements, especially in the medical field. Machine learning is an astounding field with precise outcomes in medical domains such as detection, diagnosis, imaging, personalized medicine, etc. The machine learning algorithms analyze the feature-engineered data and produce precise outcomes using different learning methods, such as supervised and unsupervised. In the case of medical applications, these algorithms play a vital role in disease diagnosis and recognition of patterns, even in the absence of medical experts. For instance, during the coronavirus (Covid-19) pandemic, machine learning algorithms have accurately identified the infected persons using chest X-ray recordings, real-time polymerase chain reaction (RT-PCR) tests, and blood samples in the early stages. However, the learning algorithms have certain limitations in recognizing an imbalanced dataset collected for the deadliest diseases such as coronary heart disease, strokes, respiratory illness, COPD, cancers, diabetes, Alzheimer's disease, TB, cirrhosis, etc. To investigate the performance of such a class-imbalanced dataset and to improve its performance in terms of false positive rates, this chapter utilizes various methods such as random oversampling, random undersampling, and SMOTE to handle the class-imbalance problem in two different medical datasets. Then, the balanced dataset is evaluated using algorithms like naïve Bayes, decision tree, and *k*-nearest neighbor in different evaluation models. The learning algorithms are evaluated with familiar metrics – precision, accuracy, recall, *F*-measure, and ROC area. Experiments on two utilized class-imbalanced datasets show that SMOTE performs better in handling class-imbalance problems.

1.1 Introduction

Machine learning (ML) is a subset of artificial intelligence (AI) that produces a precise outcome on the given feature representation without being explicitly programmed. The ML has two phases: training and testing. While training, the features are learned from the given training data, and while testing, it uses the learned features for precise outcomes. Each learning algorithm has a different learning process by its nature. Testing is also referred to as the generalization process, where the learning algorithm recognizes the new data even without learning the same during the training process [1].

Fundamentally, the learning algorithms gain information (features) from the historical data/feature representation and exchange the data to a mathematical model on training. While testing, the designed mathematical model is used to deliver solid and precise outcomes. The learning algorithm requires only the data in a suitable format for processing; it has to be transformed into proper formatting for the learning process. In general, the learning algorithm requires a data pre-processing phase that includes the removal of noise, outliers, integration, and data transformation for a suitable learning process. By and large, the ML algorithms are mainly regulated, unaided, semi-managed, or support learning. ML algorithms are classified into supervised, unsupervised, and reinforcement learning. Supervised learning can be chosen in case of a clear feature representation of data with a target class. The unsupervised learning algorithm would generally give better outcomes on feature representation without target classes, or the target classes are unknown. Reinforcement learning (RL) can be used in case sub-optimal actions are explicitly corrected. RL does not require labeled input/output pairs. Instead, it focuses on finding a balance between exploring and exploiting current knowledge. Recently, a partially supervised RL algorithm has evolved, which combines the advantages of supervised and RL algorithms.

Recently, the availability and accessibility of real-time datasets have increased the knowledge and exposure of ML algorithms to buddy research-ers [3]. Especially challenges such as Kaggle, KDDcup, etc., conducted by the enterprises and industries are one of the significant impacts in the ML research. The researchers examine the data using an ideal ML approach and make the machines learn by themselves for better outcomes. However, the imbalanced dataset has generated a large-scale volume in real-time data col-lection [4]. For example, the dataset may have two representative classes. The dataset is balanced if the two classes have an equal number of instances, say 50% each or even 60% and 40% for classes, respectively. However, this is not the case in an imbalanced dataset. One class may have 85% of instances, and another may have 15% of instances, which makes supervised learning infeasible to learn appropriately. The respective outcomes on testing will be less for all the machine learning algorithms.

Learning the imbalanced dataset using a supervised learning algorithm has certain limitations. Imbalanced class information represents a non-dis-criminate grouping of classes when the perceptions of classes are not con-veyed properly. Performance of the learning algorithm satisfies the grouping with majority class and fewer perceptions for the minority class (that has least information).

The remaining parts of this chapter have been categorized as fol-lows. Section 1.2 reviews the problem formulation on a class-imbalanced data-set. Section 1.3 discusses the models to handle the class-imbalance problem. Section 1.4 discusses the classification models used for performance evalu-ation. Section 1.5 deals with various class-imbalanced medical datasets, and the experiments have been carried out and evaluated in Section 1.6. Finally, Section 1.7 concludes the chapter.

1.2 Problem Formulation

Data collection with imbalanced class representation is pretty standard in a real-time application, especially in the medical domain. Collecting danger-ous or suspicious disease data in a large volume is critical – for example, collecting critical cancer species from a human body. Various researchers have proposed machine learning algorithms to investigate, such as imbal-anced datasets. However, the performances are not satisfactory to the mark as the medical data are real-time and need to provide higher accuracies [5].

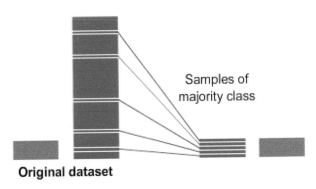

Figure 1.1   Random undersampling technique

The performance issues and limitations of class-imbalanced datasets are as follows:

- They are driven and will probably restrict the overall error to which the minority class contributes very little.

- They anticipate a comparable feature of data for all the classes.

- They acknowledge that the instances from different classes have a comparative cost.

1.3 Techniques to Handle Imbalanced Datasets

The principal goal of adjusting classes is to build the recurrence of the minority class or reduce the recurrence of the majority class to acquire a similar number of samples for both classes. There are quite a few techniques to handle the imbalance in datasets. Resampling, a data-level approach, is one of the widely adopted methods. It comprises removing samples from the majority class (undersampling) and adding additional data from the minority class (oversampling).

1.3.1 Random undersampling (RUS)

RUS is a non-heuristic technique that plans to adjust class balance through the arbitrary disposal of the majority class. The reason behind it is to adjust the dataset to reduce the complexity of the model creation. RUS strategy eliminates models from the majority class to adjust the informational collection, as shown in Figure 1.1. This strategy is appropriate for a broad scope

Oversampling

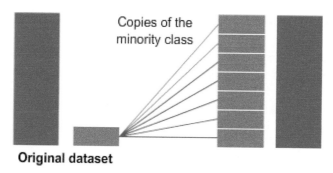

Figure 1.2 Random oversampling technique.

of applications, where the quantity of majority class models is exceptionally huge and reducing the training time decreases the learning algorithm's complexity (time and space).

Undersampling techniques are categorized into random (irregular) and informative (educational). It arbitrarily takes out models from the majority class till the information collection gets adjusted. The necessary majority class models are chosen in informative undersampling depending on a pre-specified choice measure to adjust the informational index. Exclusive strategies are proposed as pre-processing procedures for choosing informative examples for a classifier. At the time of passive choice strategies, these examples are questioned during the development interaction of the classifier. The undersampling technique may sometimes remove the instances that could be significant to improve the classification performance.

1.3.2 Random oversampling (RUS)

The random oversampling (ROS) strategy is a non-heuristic method, which works in the way that the instances of minority class are arbitrarily copied from the dataset to adjust the quantity of each dataset, as in Figure 1.2. Since oversampling does not lose the data on the complete model design, it can achieve higher performance measure accuracy. In any case, it can prompt overfitting issues and a training time since the quantity of information utilized in training the model is larger than the testing process. It is feasible to join these two sorts of inspecting strategies. Thus, it creates an arbitrary subsample of a dataset by extraction with substitution under minority class dispersion.

The oversampling technique has a computational error and time complexity on an extensive collection of imbalanced data. Generally, it is feasible to join the techniques of RUS and ROS, to deliver an arbitrary subsample of a dataset by extraction with substitution under determined class circulation. Although these methodologies have intriguing thoughts attempting to tackle the imbalanced information issue, no classifier has been developed to handle the class-imbalance data.

Besides these strategies, numerous reasonable resampling procedures have been proposed to solve the class-imbalance problem. Recently, certain experimentations have been performed against these resampling-based methodologies [6] to tackle the imbalanced information issue. In such resampling methodologies, the intelligence on testing is complex and consumes more significant complexity.

1.3.3 Synthetic minority oversampling technique (SMOTE)

SMOTE is the acronym for synthetic minority oversampling technique. Classification utilizing class-imbalanced information is one-sided for the majority class. The inclination is much more significant for high-dimensional information, where the quantity of factors surpasses the number of tests. The issue can be constricted by undersampling or oversampling, which produces class-adjusted information. Many broadened arrangement calculations depend on the customary under-testing and over-inspecting procedures.

For example, Chawla *et al.* [7] introduced SMOTE, a unique over-examining technique focusing on majority classes. Additionally, it utilizes the under-testing method to haphazardly eliminate occasions from the majority class until the extent of minority and majority classes arrive at a specific level. The analyses of Chawla *et al.* [7] have shown that the proposed approach has outperformed other state-of-the-art methods in terms of sensitivity. Given that, a few expansions of SMOTE have been proposed. In [8], the researchers have developed a model named SDC that joins a specific variation of SMOTE with a few critical calculations. The analyses have depicted that SDC has performed better compared to SMOTE.

1.4 Classification Models

1.4.1 Naive Bayes

A naive Bayes classifier is a supervised machine learning algorithm used for classification [9]. It is based on Bayes' theorem, the so-called statistical Bayesian classifier [10]. Bayes' theorem processes conditional probability, a measure of the probability of an event occurring where another event has already occurred as shown in the Figure 1.3.

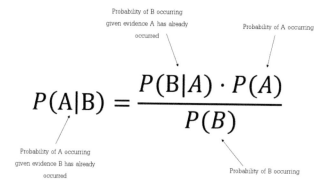

Figure 1.3 Bayes conditional probability equation.

A naive Bayes classifier is simple as its computational time for training is less. By removing the irrelevant data features, the classification performance can be improved. But the limitation of the naive Bayes classifier is that it requires a larger dataset (more number of instances or samples) to acquire better results.

1.4.2 k-Nearest neighbor classifier

The k-NN [11] is the simplest supervised machine learning technique. It assumes the similarity between the test data and the train data and classifies the test data as the most similar category of train data. This algorithm generally stores the entire data to train and classify the test data based on similarity. Thus, it has been named instance-based learning or lazy learners. It is also known as the non-parametric algorithm. That is, it does not make any assumption on any underlying data. This can also be used for regression along with the classification process.

1.4.3 Decision tree classifier

Algorithm ID3 (Iterative Dichotomiser 3) involves the java implementation of decision tree J48, developed by the WEKA group [12]. This algorithm builds decision trees specifically and uses a divide-and-conquer technique, a recursive top-down approach [13]. The internal nodes in the tree stand for the features, branches for the decision rules, and leaf nodes for the target class. It begins with the root hub, develops different branches, and builds a tree-like design. This algorithm asks a yes/no question and splits the tree into sub-trees. Information gain, gain ratio, and Gini index are used [14]. In this chapter, information gain has been used to choose the attributes at each stage. The decision tree algorithm has certain advantages over other machine learning algorithms, such as the following:

- The decision-making process is simple and easy to understand, as it follows a top-down approach.

- It is useful for solving all decision-related problems.

- It can classify the dataset with imbalanced, missing values.

- It requires less/no pre-processing of dataset compared to other classifiers.

1.4.4 Random forest

Random forest is an ensemble supervised learning strategy for classification and regression [15]. The classifier uses an ensemble of supervised classifiers, explicitly decision trees, to provide solutions for complex problems. The "forest" of the tree is constructed through the bagging or bootstrap aggregation strategy. Bagging is a group of meta-learning classifiers that works on the ensemble of the forest. The predictions are inferred by averaging the predictions of different decision trees. This ensemble algorithm consumes more computational time as it develops more trees to expand the accuracy of the result. The decision trees' depth limits the ensemble approach's performance, leading to overfitting and exactness of the solution. It has certain advantages over other machine learning classifiers, such as the following:

- more accurate than the decision tree classifier;

- effectively handles the missing data;

- produces a reasonable prediction without hyperparameter tuning;

- overcomes the overfitting problem.

1.5 Dataset Collection

For performance comparison, two different class-imbalanced datasets from the UCI repository [16–18] have been utilized.

1.5.1 Heart failure clinical records dataset

This dataset has been collected at the Faisalabad Institute of Cardiology and at the Allied Hospital in Faisalabad (Punjab, Pakistan), from April to December 2015 [17]. It comprises medical records of 299 heart failure patients of 105 women and 194 men and their ages are ranged between 40

Figure 1.4 Metadata information of heart failure clinical records dataset.

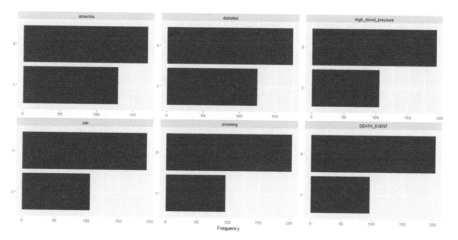

Figure 1.5 Distribution frequency of binary features – heart failure clinical records dataset.

and 95 years. This dataset contains 13 features that report clinical, body, and lifestyle information, and its metadata information is shown in Figure 1.4. Certain features are binary such as anemia, high blood pressure, diabetes, sex, and smoking, and their frequency values are shown in Figure 1.5. The discrete features (numeric values) such as age, creatinine, ejection fraction, platelets, serum creatinine, serum sodium, time, and their density values are shown in Figure 1.6. Regarding the dataset imbalance, the survived patients (death event = 0) are 203, while the dead patients (death event = 1) are 96. In statistical terms, there are 32.11% positives and 67.89% negatives, which are clearly shown in Figure 1.6 (Death_event).

1.5.2 Diabetes dataset

The diabetes dataset reviews the information about the early detection of diabetic disease among the patients of Sylhet Diabetes Hospital in Sylhet,

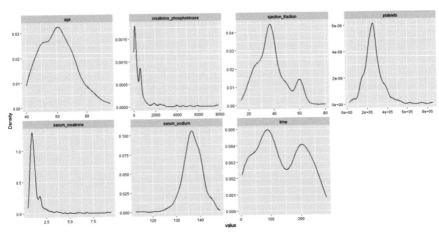

Figure 1.6   Density values of discrete features – heart failure clinical records dataset.

Figure 1.7   Metadata information of diabetes dataset.

Bangladesh [18]. The dataset has been collected through a questionnaire given to various patients and is officially approved by a doctor. It contains 16 features of which age is nominal data and the other 15 features are binaries that represent yes/no. Metadata information of this dataset is shown in Figure 1.7, nominal density distribution is shown in Figure 1.8, and frequency distribution of binary features is shown in Figure 1.9. In class-imbalanced features, the positive class of diabetes is 200 (38.41%) and the negative class of diabetes is 320 (61.53%).

1.6 Experimental Results and Discussion

To experimentally estimate the class-imbalance datasets, random oversampling, random undersampling, and SMOTE techniques have been utilized as discussed in Section 1.4. Their performances are measured using various

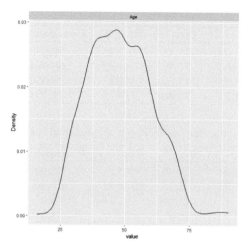

Figure 1.8 Density values of discrete features – diabetes dataset.

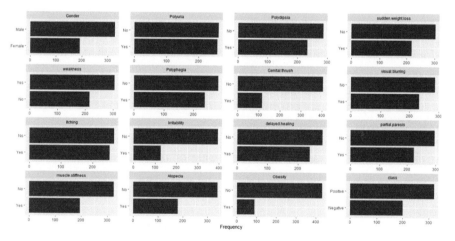

Figure 1.9 Distribution frequency of binary features – heart failure clinical records dataset.

classifiers such as naïve Bayes, decision tree, k-nearest neighbor, and random forest with familiar metrics such as precision (Pre), accuracy (Acc), recall (Rec), F-measure, and ROC area as described in the research work [19].

1.6.1 Heart failure clinical records dataset

The class-imbalanced heart failure clinical record dataset has been initially evaluated through two different evaluation methods such as 10-fold cross-validation and 70-30 training and testing, and their results are shown in

Table 1.1   Ten-fold cross-validation on imbalanced heart failure clinical records.

Classifier	Acc (%)	Pre (%)	Rec (%)	*F*-measure (%)	ROC area (%)
Naïve Bayes	76.92	71.42	46.875	56.603	85.4
Decision tree	80.60	71.59	65.625	68.474	75.1
k-NN	65.55	45.20	34.375	39.054	57.3
Random forest	81.27	70.408	71.875	71.134	88.9

Table 1.2   Training and testing on 70-30 split of imbalanced heart failure clinical records.

Classifier	Acc (%)	Pre (%)	Rec (%)	*F*-measure (%)	ROC area (%)
Naïve Bayes	74.44	78.94	44.11	56.60	78.4
Decision tree	83.33	85.18	67.64	75.40	80.9
k-NN	71.111	75.0	35.294	48.0	64.1
Random forest	83.33	88.0	64.705	74.576	90.0

Tables 1.1 and 1.2. On comparing the performance of 10-fold cross-valida-tion on the class-imbalanced dataset, the random forest has shown better per-formance in terms of precision, accuracy, recall, *F*-measure, and ROC area, as the random forest technique uses an ensemble principle for classification. Moreover, *k*-nearest neighbor has shown very bad performance compared to the other two classifiers, as it works on the principle of lazy neighbor. This is mainly due to class-imbalanced dataset values.

On comparing the performance of training and testing of 70-30 splits of a class-imbalanced dataset through Table 1.2, again random forest has shown better performance in terms of accuracy and it shares the same value as that of decision tree. However, the recall value of the decision tree is slightly greater than the performance of random forest, due to the rule-based classifi-cation principle. The other performance factors are highly induced in random forest classifiers.

To avoid such an issue, the random undersampling technique has been carried out and its performances are shown in Tables 1.3 and 1.4. In the ran-dom undersampling technique, the class-imbalanced dataset target class val-ues are reduced to 96 for both the binary class values. On comparing the performance of random undersampling, both the decision tree and the ran-dom forest have shown better values which are clearly shown in Table 1.4. However, in training and testing splits of 70-30, random forest and decision tree share certain values like the raw classification on the class-imbalanced dataset.

Further, to analyze the importance of random oversampling, the raw class-imbalanced dataset has been upgraded with random oversampling to produce the target class of 144 positive cases and 155 negative cases,

Table 1.3 Ten-fold cross-validation on imbalanced heart failure clinical records using random undersampling.

Classifier	Acc (%)	Pre (%)	Rec (%)	*F*-measure (%)	ROC area (%)
Naïve Bayes	78.125	86.48	66.66	75.294	85.4
Decision tree	80.20	79.59	81.25	80.412	79.7
k-NN	64.062	67.08	55.208	60.571	64.1
Random forest	80.20	79.591	81.25	80.41	89.4

Table 1.4 Training and testing on 70-30 split of imbalanced heart failure clinical records using random undersampling.

Classifier	Acc (%)	Pre (%)	Rec (%)	*F*-measure (%)	ROC area (%)
Naïve Bayes	77.586	87.5	67.74	76.363	84.7
Decision tree	82.758	81.818	87.096	84.375	82.8
k-NN	55.172	61.904	41.935	50.25	56.2
Random forest	82.758	88.888	77.419	82.758	89.6

Table 1.5 Ten-fold cross-validation on imbalanced heart failure clinical records using random oversampling.

Classifier	Acc (%)	Pre (%)	Rec (%)	*F*-measure (%)	ROC area (%)
Naïve Bayes	77.926	86.794	63.8889	73.6	87.7
Decision tree	87.62	86.39	88.194	87.28	90.6
k-NN	86.622	86.619	85.416	86.013	86.6
Random forest	91.973	91.095	92.361	91.729	97.3

Table 1.6 Training and testing on 70-30 split of imbalanced heart failure clinical records using random oversampling.

Classifier	Acc (%)	Pre (%)	Rec (%)	*F*-measure (%)	ROC area (%)
Naïve Bayes	84.444	94.593	74.463	83.33	89.4
Decision tree	85.55	88.636	82.978	85.714	85.9
k-NN	83.33	84.782	82.978	83.879	83.3
Random forest	94.444	93.75	95.744	94.763	96.2

respectively. Their performances are shown in Tables 1.5 and 1.6. On comparing the performances, the random forest has shown better performance on both 10-fold cross-validation and 70-30 training and testing splits. The performances are even higher compared to the random undersampling technique and it shows the efficiency of oversampling instead of the undersampling technique.

Finally, the class-imbalanced dataset has been modified using SMOTE to produce 299 records for the binary class values and their performances on classification are shown in Tables 1.7 and 1.8. On comparing the

Table 1.7   Ten-fold cross-validation on imbalanced heart failure clinical records using SMOTE.

Classifier	Acc (%)	Pre (%)	Rec (%)	*F*-measure (%)	ROC area (%)
Naïve Bayes	85.284	84.84	82.35	83.58	88.2
Decision tree	90.635	90.83	88.14	89.47	93.2
k-NN	91.973	91.095	92.361	91.729	97.3
Random forest	98.2692	96.13	99.5	97.788	98.5

Table 1.8   Training and testing on 70-30 split of imbalanced heart failure clinical records using SMOTE.

Classifier	Acc (%)	Pre (%)	Rec (%)	*F*-measure (%)	ROC area (%)
Naïve Bayes	85.55	88.636	82.978	85.714	85.9
Decision tree	94.444	93.75	95.744	94.763	96.2
k-NN	94.2307	92.64	94.02	93.333	92.6
Random forest	98.7179	97.10	100	98.523	98.9

performances, the random forest has shown better and highest performance than the other classifiers. Moreover, it has gained better performance even with undersampling and oversampling techniques.

1.6.2 Diabetes dataset

The experimentation has been repeated for the diabetes dataset to validate the performance of the class-imbalance handling mechanism using the classifiers. The initial experimentation has been carried out with class-imbalanced dataset on 10-fold cross-validation with training and testing splits of 70-30, and their results are shown in Tables 1.9 and 1.10. When the performances are compared, *k*-NN has shown better performances in terms of all metrics except the ROC area with random forest. This shows the indulgence of features present in the dataset for lazy neighboring process.

To analyze the effects of the random undersampling technique, the dataset has been modified with the process of undersampling and the binary target records are modified into 200 records each. Their classification performances are shown in Tables 1.11 and 1.12. When comparing the performance on 10-fold cross-validation, *k*-NN has shown better performance in terms of accuracy, precision, and *F*-measure than random forest, but it fails to show better performance in terms of recall and ROC area. On comparing the performance from Table 1.12 for training and testing splits, the *k*-NN has shown better performance similar to the cross-validation technique. This may be mainly due to the shuffling and indulgence of similar records on cross-validation, training, and testing.

Table 1.9   Ten-fold cross-validation on imbalanced diabetes dataset.

Classifier	Acc (%)	Pre (%)	Rec (%)	F-measure (%)	ROC area (%)
Naïve Bayes	87.8846	80.71	90.01	85.106	94.7
Decision tree	94.0385	91.2	93.5	92.345	95.4
k-NN	98.2692	96.13	99.5	97.788	98.5
Random forest	97.5	96.51	97.12	96.75	99.7

Table 1.10   Training and testing on 70-30 split of imbalanced diabetes dataset.

Classifier	Acc (%)	Pre (%)	Rec (%)	F-measure (%)	ROC area (%)
Naïve Bayes	89.1026	84.722	91.044	87.7697	95.3
Decision tree	94.2307	92.64	94.02	93.333	92.6
k-NN	98.7179	97.10	100	98.523	98.9
Random forest	98.0769	95.71	100	97.81	99.9

Table 1.11   Ten-fold cross-validation on imbalanced diabetes dataset using random undersampling.

Classifier	Acc (%)	Pre (%)	Rec (%)	F-measure (%)	ROC area (%)
Naïve Bayes	86.5	83.793	90.5	87.019	95.1
Decision tree	94.25	92.753	96.0	94.348	95.7
k-NN	96	93.396	99.1	96.116	96.0
Random forest	96.75	96.05	97.5	96.77	99.5

Table 1.12   Training and testing on 70-30 split of imbalanced diabetes dataset using random undersampling.

Classifier	Acc (%)	Pre (%)	Rec (%)	F-measure (%)	ROC area (%)
Naïve Bayes	88.33	86.44	89.473	87.930	96.6
Decision tree	95	93.22	96.49	94.82	96.3
k-NN	96.667	94.91	98.24	96.55	96.7
Random forest	95.833	93.33	98.24	95.72	99.9

To analyze the effects of oversampling, the dataset has been modified into records of 192 positive and 328 negative classes, respectively, by using random oversampling technique, and their results are shown in Tables 1.13 and 1.14. The performance of k-NN is very high compared to other classifiers on cross-validation and its shares the metrics with the decision tree during training and testing splits.

Finally, on comparing the performance of SMOTE on the class-imbalanced process of diabetes dataset as shown in Tables 1.15 and 1.16, k-NN achieves the highest performance in terms of all the metrics compared to the other classifiers and even compared to random undersampling and random oversampling techniques. This shows the efficiency of SMOTE on other class-imbalance handling mechanisms.

Table 1.13 Ten-fold cross-validation on imbalanced diabetes dataset using random oversampling.

Classifier	Acc (%)	Pre (%)	Rec (%)	F-measure (%)	ROC area (%)
Naïve Bayes	86.92	79.52	86.97	83.08	94.0
Decision tree	97.11	95.38	96.875	96.12	97.6
k-NN	99.03	98.44	98.958	98.701	99.0
Random forest	98.653	97.43	98.959	98.192	99.9

Table 1.14 Training and testing on 70-30 split of imbalanced diabetes dataset using random oversampling.

Classifier	Acc (%)	Pre (%)	Rec (%)	F-measure (%)	ROC area (%)
Naïve Bayes	83.97	75.43	79.62	77.477	92.7
Decision tree	98.71	98.14	98.141	98.141	98.6
k-NN	98.71	98.14	98.141	98.141	98.6
Random forest	98.07	96.363	98.141	97.247	99.9

Table 1.15 Ten-fold cross-validation on imbalanced diabetes dataset using SMOTE.

Classifier	Acc (%)	Pre (%)	Rec (%)	F-measure (%)	ROC area (%)
Naïve Bayes	88.946	83.58	92.33	88.326	96.8
Decision tree	96.0385	95.2	96.7	95.455	97.4
k-NN	99.892	98.13	99.5	98.889	99.5
Random forest	99.5	97.65	98.12	97.75	99.7

Table 1.16 Training and testing on 70-30 split of imbalanced diabetes dataset using SMOTE.

Classifier	Acc (%)	Pre (%)	Rec (%)	F-measure (%)	ROC area (%)
Naïve Bayes	89.13	88.32	90.121	89.310	96.8
Decision tree	95.89	94.12	97.33	98.82	97.3
k-NN	98.87	96.91	99.24	97.66	98.7
Random forest	98.833	95.33	99.12	96.72	99.9

1.7 Conclusion

Class-imbalance problems are precision drive and the expenses on classification in terms of any metrics have no proper appropriation. This kind of class imbalance is majorly seen in medical records, as it is difficult to simulate or record the negative cases such as cancer, Covid-19, etc. To deal with such a class-imbalance problem, this chapter provides various class-imbalance handling mechanisms such as random undersampling, random oversampling, and SMOTE, and their performances are evaluated using classifiers such as naïve Bayes, k-NN, decision tree, and random forest. Their performances are estimated using familiar metrics such as accuracy, precision, recall, F-measure, and ROC area. Through the experimentation on a two-class-imbalanced dataset, it is clear that SMOTE has a better class-imbalanced handling mechanism over the random undersampling and oversampling approaches.

References

[1] Jordan, M. I., & Mitchell, T. M. (2015). Machine learning: Trends, perspectives, and prospects. Science, 349(6245), 255–260.

[2] Ahn, G., Park, Y. J., & Hur, S. (2020). A membership probability–based undersampling algorithm for imbalanced data. Journal of Classification, 1–14.

[3] Ray, S. (2019, February). A quick review of machine learning algorithms. In 2019 International conference on machine learning, big data, cloud and parallel computing (COMITCon) (pp. 35–39). IEEE.

[4] Aridas, C. K., Karlos, S., Kanas, V. G., Fazakis, N., & Kotsiantis, S. B. (2019). Uncertainty based under-sampling for learning naive Bayes classifiers under imbalanced data sets. IEEE Access, 8, 2122–2133.

[5] Brownlee, J. (2020). Smote for imbalanced classification with python. Machine Learning Mastery, 16.

[6] Goswami, T., & Roy, U. B. (2021). Classification Accuracy Comparison for Imbalanced Datasets with Its Balanced Counterparts Obtained by Different Sampling Techniques. In ICCCE 2020 (pp. 45–54). Springer, Singapore.

[7] Chawla, N. V., Bowyer, K. W., Hall, L. O., & Kegelmeyer, W. P. (2002). SMOTE: synthetic minority over-sampling technique. Journal of artificial intelligence research, 16, 321–357.

[8] Akbani, R., Kwek, S., & Japkowicz, N. (2004, September). Applying support vector machines to imbalanced datasets. In European conference on machine learning (pp. 39–50). Springer, Berlin, Heidelberg.

[9] Padmavathi, S., & Ramanujam, E. (2015). Naïve bayes classifier for ecg abnormalities using multivariate maximal time series motif. Procedia Computer Science, 47, 222–228.

[10] Nikam, S. S. (2015). A comparative study of classification techniques in data mining algorithms. Oriental journal of computer science & technology, 8(1), 13–19.

[11] Dudani, S. A. (1976). The distance-weighted k-nearest-neighbor rule. IEEE Transactions on Systems, Man, and Cybernetics, (4), 325–327.

[12] Vasudevan, P. (2014). Iterative dichotomiser-3 algorithm in data mining applied to diabetes database. Journal of Computer Science, 10(7), 1151.

[13] Mathuria, M. (2013). Decision tree analysis on j48 algorithm for data mining. Intrenational Journal of Advanced Research in Computer Science and Software Engineering, 3(6).

[14] Jadhav, S. D., & Channe, H. P. (2016). Comparative study of K-NN, naive Bayes and decision tree classification techniques. International Journal of Science and Research (IJSR), 5(1), 1842–1845.

[15] Pal, M. (2005). Random forest classifier for remote sensing classification. International journal of remote sensing, 26(1), 217–222.

[16] Asuncion, A., & Newman, D. (2007). UCI machine learning repository.

[17] Chicco, D., & Jurman, G. (2020). Machine learning can predict survival of patients with heart failure from serum creatinine and ejection fraction alone. BMC medical informatics and decision making, 20(1), 1–16.

[18] Islam, M. F., Ferdousi, R., Rahman, S., & Bushra, H. Y. (2020). Likelihood prediction of diabetes at early stage using data mining techniques. In Computer Vision and Machine Intelligence in Medical Image Analysis (pp. 113–125). Springer, Singapore.

[19] Ramanujam, E., Padmavathi, S., Baskar, L. A., & Niwin, P. (2021). Ensemble Feature Selection for the Recognition of Human Activities and Postural Transitions on Smartphones. International Journal of Service Science, Management, Engineering, and Technology (IJSSMET), 12(5), 80–101.

2

Improving Heart Disease Diagnosis using Modified Dynamic Adaptive PSO (MDAPSO)

J. Shanthalakshmi Revathy[1], J. V. Anchitaalagammai[2], and S. Hariharasitaraman[3]

[1]Department of Computer Science and Engineering, Velammal College of Engineering and Technology, India.
[2]Department of Computer Science and Engineering, Velammal College of Engineering and Technology, India.
[3]School of Computing Science and Engineering, VIT Bhopal University, India.
Email: jslr@vcet.ac.in; jva@vcet.ac.in; hariharasitaraman@gmail.com

Abstract

This study aims to expand an automatic heart disease diagnosis system to categorize the danger of heart diseases. There are several things to consider while examining a patient's heart condition, and it is not an easy process, which makes the physician's work difficult. However, the professionals would like a precise tool that identifies the risk factors connected with heart disease. Particle swarm optimization (PSO) is a process for explaining sophisticated and complex issues that are difficult to address using traditional approaches. By developing modified dynamic adaptive particle swarm optimization (MDAPSO), we use weighted PSO to evaluate threshold values for support and confidence. After the data is converted into binary numbers, the PSO algorithm searches for each feature's most favorable fitness value and then determines comparable support and confidence as minimum threshold values. The authors apply MDAPSO for feature selection. For feature selection, we use MDAPSO. In the present research, the heart disease dataset was acquired from the UCI (machine learning repository).

2.1 Introduction

One of the most critical concerns in recent years has been medical data mining, which relies on analysis and statistical reasoning, machine learning techniques, and pattern recognition to discover relations and unseen samples in the datasets of the patients. Generally, every activity in medication can be divided into six areas: screening, diagnosis, treatment, prognosis, monitoring, and management. The accuracy and sensitivity have meticulous significance in the analysis and prediction of diseases. Its optimistic response can influence the doctor's analysis to speed up the process of diagnosis and prognosis. Therefore, the costs of treatment can be reduced, and the rate of health in society can be increased. In the real world, medical databases are usually filled with irrelevant and redundant features that increase the database's dimension or lead them to a curse of dimensionality. Then, the accuracy, computational cost, and speed of the learning process are affected. Dimensionality reduction techniques have been anticipated to resolve this problem. The challenge of feature subset selection entails finding and choosing significant features from a wider set of frequently redundant, uninteresting, and different features. Here, we concentrate our study on the important issue of human heart disease. In India, the disease related to the heart is the foremost origin of mortality.

As per the data of the World Health Organization, each year, 12 million people pass away due to heart illness. In the United States, one person dies every 36 seconds due to various types of heart disease. Most of the people died of heart disease catastrophe only. Heart illnesses include heart disease related to coronary, cardiomyopathy disease, and cardiovascular disease, among others. The illness concerning cardiovascular disease affects blood flow throughout the body and the blood arteries that link the heart to the rest of the body. So we strive to reduce heart diseases. The cardiovascular disease leads to coronary artery disease, high blood stroke, pressure, etc. Serious effects of cardiovascular disease lead to death. It is a kind of coronary artery illness in which the heart's blood and oxygen supply are inadequate due to a narrowing of the coronary arteries. Heart attacks and chest discomfort are other symptoms of coronary heart disease. Medical judgment is not a simple task that requires operating precisely and professionally. Reasons like high blood pressure, smoking, and family history add to increased heart disease. High cholesterol levels, hyper stiffness, poor nutrition, and other factors can contribute to heart diseases.

The following are some of the most common symptoms of various forms of heart illness:

- peripheral artery disease;

- stroke;

- high BP;

- fatigue;

- cardiac arrest;

- palpitation;

- congestive heart failure;

- pain in chest;

- high cholesterol;

- arrhythmia;

- peripheral artery disease.

The researchers proclaimed many automatic heart disease diagnosis solutions. When a heart attack is identified, detection speed is necessary to save the patient's life and avert heart damage. The standard PSO has many limitations that are not yet analyzed. This corresponds to sensitivity to transformations such as translation, rotation, scale, local convergence, stability, and first hitting occasion. As long as these restrictions persist, PSO's performance is not scalable to a broad series of optimization problems. In this regard, the current study's goal was to aid clinicians in identifying cardiac disease. The following assertions summarize the uniqueness of this research.

   i. To maintain steadiness amid exploitation and exploration, inertia weight is important. The inertia weight determines the involvement pace of a particle's speed at the existing time step to its prior velocity.

   ii. Larger and smaller inertia weights make global and local searches easier.

   iii. The introduction of dynamical inertia weight adjustment is used to improve the capabilities of PSO.

   iv. The basic components of the PSO are modified for improving performance. Basic components are velocity clamping, inertia term, and acceleration coefficients. Multiple changes are made to more than one of the PSO's core components.

   v. For future implementation, a validation developed model is used.

The following is the sequence in which the remaining sections of this article should be read. Section 1.2 shows prior work on the heart disease detection system as well as a nature-inspired algorithm and its modification. Section 1.3 explores the proposed priority-based weighted PSO model assisted with the modified dynamic adaptive particle swarm optimization (MDAPSO). Section 1.4 exhibits the performance results and their verification. The MDAPSO model's efficacy is summarized in the last section, along with recommendations for further study.

2.2 Background and Related Work

This section concerns the existing ideas used in heart disease prediction. The dataset used in various research articles and the information related to heart disease is extracted. Particularly, it highlights how the risk of the disease is classified. The conventional approaches for feature selection, feature extraction, illness classification, disease clustering, and disease severity analysis are also examined in the related study.

2.2.1 PSO modification

PSO has been used in many applications with modifications to improve the performance. The major alterations to the basic PSO are directed to improve two components. The first component is the convergence of the PSO, and the second component increases the group's diversity. The modification is mainly classified into PSO categories: field searching space addition, the parameter fine-tuning, and fusion with other techniques. Velocity constriction, inertia weight, velocity clamping, various methods of affecting the personal finest and worldwide locations, cognitive and social coefficient, and various velocity models are among the factors that may be changed. The modified PSO is based on adjusting Inertia weight in context to trial and error technique. Appropriate choice of the inertia weight gives stability involving local and global searching. These concepts proposed a linearly diminishing, linearly growing, and sigmoid diminishing inertia weight to get better PSO performance. There are improvements between the three techniques: sigmoid diminishing inertia is close to the most favorable result superior to the others, and linearly increasing weight has fast convergence capability better than the others.

2.2.2 Feature selection

The main process of developing an effective prediction model is feature selection. In the heart disease diagnosis, every factor's weight cannot be measured as static because the influence of factors differs for everybody. Paul *et al.* [1] applied the fuzzy logic concept for the heart disease diagnosis. It consisted of three key steps, i.e., fuzzification step, rule-based step, and defuzzification step. Here, defuzzification uses the centroid technique. This method has 13 participation features and 1 production feature. It justified its work with the highest accuracy of 93.33%. Zamani and Nadimi-Shahraki [2] proposed whale optimization for feature selection. By removing dissimilar and duplicate characteristics, it attempted to reduce the amount of the data and increase the effectiveness of learning algorithms. It has three main processes namely encircling prey, spiral bubble-net attacking, and searching for prey. Iftikhar *et al.* [3] suggested a hybrid strategy for classifying cardiac illnesses and identifying risk factors linked with them. They utilized SVM classifiers, GA, and PSO optimization methods to classify various heart disease scenarios. The fundamental purpose of the GA and PSO algorithms is to use fewer but more discriminative features to improve SVM classification accuracy dramatically. The GA and PSO are population-based evolutionary algorithms that initiate the similarly condensed most favorable feature set and best concluding point.

2.2.3 Classification and clustering

Indecision about unidentified with many alternatives is one of the unidentified illness characteristics. Irrespective of various subtypes of that illness, there may be very little feature function. Clustering is a well-known method in data mining that calculates the similarity categorization to identify illness subtypes. Shen *et al.* [4] described an image-based automated segmentation and categorization of skin lesions. SVM and *k*-NN classifiers are used to extract picture characteristics and diagnose illness. Dangare and Apte propose that the techniques related to data mining categorization have been used to increase the research of cardiac illness prediction systems [5]. It uses the multilayer perceptron neural network (MLPNN). Previous studies may be executed on likely patients to obtain additional safety measures to decrease the impact of having such a condition [6], and this study proposes consistent ways to forecast cardiac illness at a young age, which is crucial for lives. A variety of machine learning (ML) classification algorithms, including naive

Bayes, decision tree J48, support vector machine (SVM), *k*-nearest neighbor (K-NN), stochastic gradient descents (SGD), Adaboost, JRip, and others, were used for the categorization and forecasting of disease related to heart, with promising results [7].

Kelwade and Salankar [8] offered a computer-aided decision support system that exhibited a decrease in the forecasting period for the dataset of heart disease, while Anooj [9] proposed organized learning methods for the prediction of the heart disease dataset. Particle swarm optimization was used to produce notable values for cardiac disease in [10]. Fatima and Pasha [11] presented a good classification accuracy for heart disease in the type of a relative investigation of diverse machine learning algorithms for the judgment of heart disease as a review study. This demonstrated the appropriateness of tools and machine learning algorithms to be employed for heart disease decision-making and analysis.

There is much relevant work in heart disease diagnostic prediction to develop better optimum categorization techniques. Various issues might arise in an impoverished illness classification scheme, highlighting the importance of information specialist policies. There are several difficulties in weighing health responses and dealing with ambiguity. The present study tries to figure out how characteristics and symptoms are linked to category classification.

2.3 Proposed Approach

The proposed feature selection and categorization methods are argued in this section. The flowchart that explains the operation flow in the suggested technique is revealed in Figure 2.1.

2.3.1 PSO algorithm

At each phase, the PSO technique primarily aids in altering the velocity of every element toward its pbest and gbest places. A random term is used to provide weight to velocity. The random numbers produced for velocity are based on the location of pbest and gbest.

The standard steps of PSO are given as follows:

1. Initially, random positions and velocities are assigned to the population of particles with D dimensions in the given issue space.

2. For every particle, calculate the required fitness function of optimization with d variables.

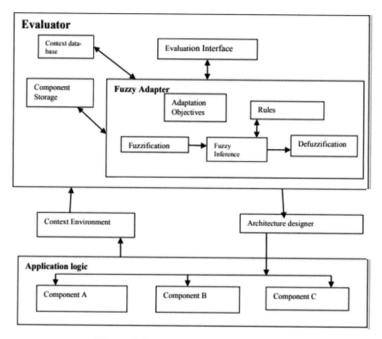

Figure 2.1 Detailed flow of operation.

3. The fitness evaluation of every particle is compared with the particle's pbest. If the existing assessment is improved than the pbest, then set pbest position equivalent to the existing place in D-dimensional space.

4. The strength assessment is compared with the previous best. If the existing assessment is improved than gbest, rearrange gbest to the current array index value of particle.

5. The particle speed and place are modified according to the following equations:

$$V_i^{k+1} = V_i^k + C_i^* \, rand(.) * (pbest - X_i^k) + C_2^* \, rand \, (.) * (gbest - X_i^k) \quad (2.1)$$

$$X_i^{k+1} = x_i^k + v_i^{k+1} \quad (2.2)$$

v_i^k is velocity vector and v_i^{k+1} is the customized velocity and is the positioning vector of particle i at generation k. C_1 and C_2 represent the cognitive and social coefficients, correspondingly, that control particle speed.

6. Continue to Step 2 until the satisfactorily excellent strength or the utmost number of iterations (generations) is met.

The V_{max} is the highest velocity that may be used to regulate a global inspection feature of particle swarm. Exploration refers to the ability to test several parts of the issue space to find a high-quality optimal. Particles may fly over superior solutions and facilitate global exploration if V_{max} is too high or too low, while particles may not seek adequately beyond local favorable regions and help narrow utilization if V_{max} is too small or too high. Exploitation refers to the capacity to narrow the search to a promising candidate solution to determine the optimum accurately. When restricted exploitation occurs, they may become stuck in local optima, not capable of going far adequate in the issue space to improve their position. PSO performance can be improved by balancing exploration and exploitation penetration procedures. This exploration and exploitation trade-off is influenced by modifying and tuning some parameters, namely current motion, inertia weight, and cognitive and social coefficients.

2.3.2 Inertia weight

The inertia weight model was created to help with exploitation and exploration management. The inertia weight was capable of domineering the exploration and exploitation methods and removing the need for V_{max}. The PSO algorithm by adding an inertia weight was mainly published in 1998. Although the inertia weight successfully fulfilled the primary goal, it could not wholly abolish the requirement for velocity clamping. The inertia weight (w) organizes the momentum of the particle by weighing the part of the earlier speed. The velocity and position update with an inertia weight was composed in eqn (2.3) and (2.4). It can be identified that the equations are similar to eqn (2.1) and (2.2), and utilizing the totaling of the inertia weight was an increasing feature of V_i^k in eqn (2.3)

$$V_i^{k+1} = w*V_i^k + C_i^* \, rand(.)*(pbest - X_i^k) + C_2^* \, rand \, (.)* \, (gbest - X_i^k) \qquad (2.3)$$

$$X_i^{k+1} = X_i^k + V_i^{k+1} \qquad (2.4)$$

2.3.3 Modified dynamic adaptive particle swarm optimization (MDAPSO)

MDAPSO was created to clarify the PSO premature convergence issue linked with common multi-peak, high-dimensional operating optimization issues to enhance convergence speed and global optimum. The deviation was fine-tuned using a dynamic adaptive method based on congregate degree and current swarm diversity and the collision's impact on the explore performance of the swarm. Eqn 3.5 depicts the usage of inertia weight calculation in the algorithm.

$$\omega_t = \omega_{min} + (\omega_{max} - \omega_{min}) \times F_t \times \varphi_t , \tag{2.5}$$

where ω_{min} and ω_{max} are the least and utmost inertia weight values, and is F_t the present number of iterations, φ_t is the weight adjustment operation, respectively:

$$F_t = 1 - \frac{2}{\pi} arc\tan(E) \tag{2.6}$$

where E is the fitness of the group.

$$\varphi_t = e^{(-t^{2/(2\sigma^2)})} \tag{2.7}$$

where $\sigma = \frac{T}{3}$ and T are the total numbers of iterations:

$$E = \frac{1}{N} \sum_{i=1}^{N} (f(x_i) - f_{avg})^2 , \tag{2.8}$$

where N is the swarm size, $f(x_i)$ is the fitness of particle i, and f_{avg} is the current average fitness of the swarm:

$$f_{avg} = \frac{1}{N} \sum_{i=1}^{N} f(x_i). \tag{2.9}$$

2.4 Experimental Setup and Results

2.4.1 Dataset

Although the dataset has 76 characteristics, most published research only uses a subset of 14 of them as given in Table 2.1. The authors utilize the Cleveland dataset, which machine learning researchers have used till now. The goal field denotes the patient's occurrence of heart illness. It is a number that ranges from 0 (no value) to 4. Tests with the Cleveland dataset have concentrated on distinguishing between presence (values 1, 2, 3, and 4) and absence (values 0, 1, 2, 3, and 4).

We put our MDAPSO classification algorithm to the test for heart disease diagnosis, and we assessed performance factors independently. The suggested method and other data mining algorithms were evaluated using identical training and test datasets.

Table 2.1   List of features after feature extraction.

Name	Type	Explanation
Age	Continuous	Age in years
Sex	Discrete	1 = male; 0 = female
CP	Discrete	Chest pain type: 1 = typical angina; 2 = atypical angina 3 = non-angina pain; 4 = asymptomatic
Trestbps	Continuous	Resting blood pressure in (mm Hg)
Chol	Continuous	Resting blood pressure in (mm Hg)
Fbs	Discrete	Serum cholesterol in (mg/dl) FBS > 120 (mg/dl) 1 = yes; 0 = no
Respect	Discrete	Resting electrocardiographic results: 0 = normal 1 = having ST-T wave abnormality 2 = showing probable or defined left ventricular hypertrophy
Thalach	Continuous	Maximum heart rate achieved
Exang	Discrete	Exercise-induced angina: 1 = yes; 0 = no
Oldpeak ST	Continuous	Depression induced by exercise relative to rest
Slope	Discrete	The slope of the peak exercise segment: 1 = up-sloping; 2 = flat; 3 = down-sloping
ca	Discrete	Number of major vessels colored by fluoroscopy that range between 0 and 3
Thal	Discrete	3 = normal; 6 = fixed defect; 7 = reversible defect
Diagnosis	Discrete	Diagnosis classes: 1 = healthy 3 = patient who is subject to possible heart disease

2.4.2 Performance metrics

To determine the efficacy of the suggested algorithm, performance measures like precision, recall, accuracy, and specificity are used. The formulae below can be used to calculate the above-mentioned performance measures. The amount of correct heart disease predictions divided by the number of forecasted cases is known as accuracy. Precision (P) is the percentage of correctly anticipated affirmative cases. The recall rate, also recognized as the true positive (TP) rate, is the percentage of positive instances that were rightly acknowledged. The true negative (TN) rate is the negative cases' percentage that was correctly categorized. The formula for performance metrics is shown in Table 2.2.

Table 2.2 Performance metrics.

Accuracy	$\dfrac{TP + TN}{TP + TN + FP + FN}$
Precision	$\dfrac{TP}{TP + FP}$
Recall	$\dfrac{TP}{FN + TP}$
TNR	$\dfrac{TN}{FP + TN}$

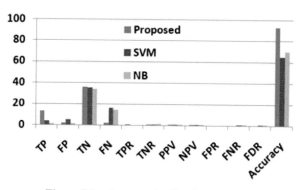

Figure 2.2 Accuracy in Cleveland dataset.

2.4.3 Result

With 76 characteristics and 303 data, the modified dynamic adaptive parti-cle swarm optimization (MDAPSO) algorithm was used to diagnose heart illness. Data mining tools are employed to assess the various performance indicators.

This section compares our proposed MDAPSO method to existing gen-eral classification algorithms such as SVM and NB algorithms in terms of sensitivity, accuracy, specificity, and precision in various datasets.

Figure 2.2 illustrates the accuracy of the classification results of the Cleveland dataset. MDAPSO specifies the heart disease diagnosis, which consists of a fully heart-related clinical dataset. Among the 303 data records, MDAPSO provides better results with high accuracy. The implementation result reveals that MDAPSO has a 92.452 percentage of accuracy, SVM has

Table 2.3   Cleveland dataset.

Dataset 1	Cleveland		
	MDAPSO	**SVM**	**NB**
TP	13	4	1
FP	2	5	1
TN	36	35	34
FN	2	16	14
TPR	0.866	0.2	0.066
TNR	0.947	0.875	0.971
PPV	0.866	0.444	0.5
NPV	0.947	0.686	0.708
FPR	0.0526	0.125	0.028
FNR	0.133	0.8	0.933
FDR	0.133	0.55	0.5
Accuracy	92.452	65	70

Table 2.4   Hungarian dataset.

Dataset 2	Hungarian		
	MDAPSO	**SVM**	**NB**
TP	26	18	18
FP	1	4	5
TN	34	31	30
FN	2	10	10
TPR	0.928	0.6428	0.642
TNR	0.971	0.885	0.857
PPV	0.962	0.818	0.782
NPV	0.944	0.756	0.75
FPR	0.028	0.114	0.142
FNR	0.071	0.357	0.357
FDR	0.037	0.181	0.217
Accuracy	95.238	77.77	76.19

a 65 percentage of accuracy, and NB has a 70% accuracy. The achieved accuracy demonstrates the MDAPSO algorithm's efficacy.

Table 2.3 illustrates the precision values for measuring up to the algorithms. The ultimate performance rating is based on the average of these individual values. MDAPSO – 0.0526, SVM – 0.8, and NB – 0.933 are the precision findings shown in Table 2.4. The findings demonstrate the MPDAPSO algorithm's reliability. To assume the accuracy values of the MDAPSO algorithm, we can simply compare the produced outcomes to the pre-identified clinical laboratory findings.

When categorizing positive records, the superior in classifying algorithm's sensitivity should be strong. To be classified as having heart disease,

Table 2.5 Switzerland dataset.

Dataset 3	Switzerland		
	MDAPSO	**SVM**	**NB**
TP	7	7	5
FP	2	13	12
TN	1	1	1
FN	1	1	1
TPR	0.875	0.875	0.833
TNR	0.33	0.071	0.076
PPV	0.777	0.35	0.294
NPV	0.5	0.5	0.5
FPR	0.66	0.928	0.923
FNR	0.125	0.125	0.166
FDR	0.222	0.65	0.705
Accuracy	72.72	36.36	31.57

Table 2.6 VA dataset.

Dataset 4	VA data		
	MDAPSO	**SVM**	**NB**
TP	14	3	5
FP	2	8	3
TN	14	5	2
FN	3	3	3
TPR	0.823	0.5	0.625
TNR	0.875	0.384	0.4
PPV	0.875	0.272	0.625
NPV	0.823	0.625	0.4
FPR	0.125	0.615	0.6
FNR	0.176	0.5	0.375
FDR	0.125	0.727	0.375
Accuracy	84.84	42.105	53.846

several characteristics must be present. The sensitivity measure successfully decreases the misclassification outcomes. Table 2.5 indicates the sensitivity values of the MDAPSO algorithm. It attained 0.866, 0.2, and 0.066 sensitivity values for the sequential prediction. For those specific categories, the proposed MDAPSO algorithm produced higher results.

The algorithm's specificity is strong; thus, it will be effective at identifying negative records. The algorithm's performance is severely harmed by erroneous recognition. Table 2.6 shows the specificity values for the MDAPSO algorithm classification as 0.947, SVM − 0.875, and NB − 0.971. The attained result shows the good domination of the algorithm.

To compare the efficiency of the MDAPSO algorithm, the accuracy parameter quantifies as first was considered. The proposed algorithm achieves an accuracy value of 95.238 Hungarian dataset, 72.72 in Switzerland dataset, and 84.84 in the VA dataset. This is a better result when compared with other algorithms.

2.5 Conclusion

The estimation of support and confidence is done automatically, and no manual processing is required. The population selection for the PSO is based on the user selection because by this, users can test the details of the output based on the results generated for the various population values. The authors can set its value to a fixed value. The modified dynamic weighted PSO (MDAPSO) method was used in this work to assess these two metrics of support and confidence rapidly and objectively. From the graphs, it is interpreted that the weighted PSO based on the algorithm MDAPSO increases the support and confidence, and the run time taken is reduced considerably. As a result, by using alternative datasets, mining performance for big databases may be improved. Further implementation of the modified dynamic adaptive particle swarm optimization (MDAPSO) will generate the results that will be analyzed and compared with the results of the non-weighted PSO.

References

[1] A. K. Paul, P. C. Shill, M. R. I. Rabin, A. Kundu, and M. A. H. Akhand, "Fuzzy membership function generation using DMS-PSO for the diagnosis of heart disease," in 2015 18th International Conference on Computer and Information Technology (ICCIT), Dec. 2015, pp. 456–461, doi: 10.1109/ICCITechn.2015.7488114.

[2] H. Zamani and M.-H. Nadimi-Shahraki, "Feature Selection Based on Whale Optimization Algorithm for Diseases Diagnosis," Int. J. Comput. Sci. Inf. Secure., vol. 14, no. 9, pp. 1243–1247, 2016.

[3] S. Iftikhar, K. Fatima, A. Rehman, A. S. Almazyad, and T. Saba, "An evolution-based hybrid approach for heart diseases classification and associated risk factors identification," Biomed. Res., vol. 28, no. 8, pp. 3451–3455, 2017.

[4] J. Shen et al., "Increased prevalence of coronary plaque in patients with psoriatic arthritis without prior diagnosis of coronary artery disease," Ann. Rheum. Dis., vol. 76, no. 7, pp. 1237–1244, Jul. 2017, DOI: 10.1136/annrheumdis-2016-210390.

[5] C. S.Dangare and S. S.Apte, "Improved Study of Heart Disease Prediction System using Data Mining Classification Techniques," Int. J. Comput. Appl., vol. 47, no. 10, pp. 44–48, Jun. 2012, doi: 10.5120/7228-0076.

[6] Y. Khourdifi and M. Bahaj, "Heart Disease Prediction and Classification Using Machine Learning Algorithms Optimized by Particle Swarm Optimization and Ant Colony Optimization," Int. J. Intell. Eng. Syst., vol. 12, no. 1, pp. 242–252, Feb. 2019, doi: 10.22266/ijies2019.0228.24.

[7] S. Mohan, C. Thirumalai, and G. Srivastava, "Effective Heart Disease Prediction Using Hybrid Machine Learning Techniques," IEEE Access, vol. 7, pp. 81542–81554, 2019, DOI: 10.1109/ACCESS.2019.2923707.

[8] J. P. Kelwade and S. S. Salankar, "Radial basis function neural network for prediction of cardiac arrhythmias based on heart rate time series," in 2016 IEEE First International Conference on Control, Measurement and Instrumentation (CMI), Jan. 2016, pp. 454–458, DOI: 10.1109/ CMI.2016.7413789.

[9] P. K. Anooj, "Clinical decision support system: Risk level prediction of heart disease using weighted fuzzy rules," J. King Saud Univ. - Comput. Inf. Sci., vol. 24, no. 1, pp. 27–40, Jan. 2012, doi: 10.1016/j. jksuci.2011.09.002.

[10] V. Krishnaiah, G. Narsimha, and N. S. Chandra, "Heart Disease Prediction System Using Data Mining Technique by Fuzzy K-NN Approach," 2015, pp. 371–384.

[11] M. Fatima and M. Pasha, "Survey of Machine Learning Algorithms for Disease Diagnostic," J. Intell. Learn. Syst. Appl., vol. 09, no. 01, pp. 1–16, 2017, DOI: 10.4236/jilsa.2017.91001.

[12] Zhu, H., Hu, Y., & Zhu, W., "A dynamic adaptive particle swarm optimization and genetic algorithm for different constrained engineering design optimization problems," Adv. in Mech. Eng.2019, DOI: 10.1177/1687814018824930

[13] AlirezaALFI, "PSO with Adaptive Mutation and Inertia Weight and Its Application in Parameter Estimation of Dynamic Systems," Acta Automatica Sinica, Volume 37, Issue 5, May 2011, pp. 541–549

3

Efficient Diagnosis and ICU Patient Monitoring Model

Premanand Ghadekar, Pradnya Katariya, Shashank Prasad, Aishwarya Chandak, Aayush Agarwal, and Anupama Choughule

Vishwakarma Institute of Technology, India
Email: ppghadekar@gmail.com; pradnya.katariya17@vit.edu;
shashank.prasad17@vit.edu; aishwarya.chandak17@vit.edu;
aayush.agarwal17@vit.edu; anupama.choughule17@vit.edu

Abstract

Healthcare is a sector that is expeditiously developing in technology and services. In recent years, the Covid-19 pandemic has drastically affected the working of the healthcare sector; people are apprehended to visit hospitals for any treatment. But evolution in modern technologies has opened multiple paths to improve and modernize the working of the healthcare sector. The proposed system is a multi-layered disease prediction model that analyzes numerous factors for predicting diseases. The system analyzes the symptoms using a modified decision tree algorithm that predicts the possible illness and suggests the test accordingly. The model is trained individually for each type of test format. For image type, reports were classified with convolutional neural networks. For PDF type, the data was extracted using optical character recognition (OCR). The model uses the Levenshtein distance to find unigrams and bigrams. The match is further analyzed, and a detailed summary of the report gets generated. Report summary and the predicted disease are provided to the patient with the list of home remedies. Further, a specialized doctor receives all the medical diagnosis details when a patient books an appointment. Hospitals usually face the problem of patient versus nurse ratio. It creates management issues to the critical ward. Patients are left unattended and can cause death threats. The proposed system analyzes multiple and dynamic factors. It increases the accuracy of the prediction. The proposed hospital monitoring system observes the vital signs on the patient monitors beside the ICU beds and notifies the hospital staff after encountering the abnormality. The model dynamically calculates the threshold value for each vital sign considering multiple factors like age, gender, and medical history of the patient. By understanding the patient's current medical condition, the model responds to change in vital signs and gives an idea about the organ's condition. Machine learning algorithm – random forest regression helps in calculating the threshold values of heart rate (HR) and respiratory rate (RR). Equations for blood pressure (BP) get the threshold values depending on age and gender. These custom thresholds for specific patients reduce false alarms, which was a significant concern in the previous monitoring system.

3.1 Introduction

Disease prediction is a way to recognize patient health by applying different techniques on patient treatment history considering multiple factors. Prediction of a disease by using a patient's symptoms and history has been striving for a decade. Symptoms are a fundamental criterion while predicting any illness; as multiple diseases show similar symptoms, it may lead to predicting false positives. As far as patient care is concerned, accuracy is the most important factor. Hence, the quality of the system improves when it is computerized. The proposed system enables multiple layers while predicting from a set of diseases to avoid any false alarms. The system initially analyzes the patient's symptoms and provides a set of medical tests that a patient should undergo. After the patient's test completion, the prediction model analyzes the medical test reports to predict the disease with higher accuracy.

The core objective of any hospital is making sure that a patient receives the optimal care and treatment. In a general hospital, patients are present in multiple locations. In this situation, monitoring system is beneficial in the increase in productivity. Nurses do not have to run from one room to another to ensure that everything is fine with the patients. Nurses' attention toward the patients can be utilized completely by reducing their workload. Hence, they can attend more patients, which help in the growth of the hospital. The proposed system overcomes the problem of a low doctor-to-patient ratio.

Computer vision is an interdisciplinary field that works on how a computer gains high-level understanding from digital images, which helps in extracting information from images and different file formats. The hospital monitoring system uses computer vision to extract vital signs data from the side monitor display screen. The system further understands the extracted data and sends an alert to the responsible hospital staff if necessary.

3.2 Main Text

The system consists of two parts – disease prediction and hospital monitoring system.

3.2.1 Disease prediction

The proposed system has a multi-layered structure that improves the accuracy of the prediction.

a) Analyzing symptoms:

The algorithm initially asks the user to select the observed symptoms. The selected symptoms are analyzed. A JSON file contains the following data:

 i. disease name;

 ii. symptoms associated with each disease;

 iii. test name for each disease;

 iv. parameters to check-in test reports and their standard values.

The proposed system consists of a list of diseases with the corresponding sets of symptoms; the input list of symptoms received from the patient (Figure 3.1) should be a subset of the symptom sets created for each disease. Standard decision trees use an approach to check such conditions that help to develop clean or pure subtrees, i.e., subtrees that have nodes of the same class. However, in the current scenario, the model consists of seven classes of diseases and at least 50 symptoms, thus creating a decision tree with conditions at each node to check whether a particular symptom is present in the input list to create a pure subtree that will generate more than 100 nodes, adding to the computation.

A modified algorithm is created, which takes decisions based on the entered symptoms and not all possible symptoms. Entered symptoms are checked, and a list of possible diseases is generated. With every symptom, the list of diseases increases with all the possible diseases for that particular symptom. To avoid redundancy, the model contains an extra condition that if three or more symptoms match, then include the disease for further process. The algorithm is extra careful about the chances of Covid-19 and includes it in the possible diseases list, even if there is a two-symptom match. The algorithm performs an additional

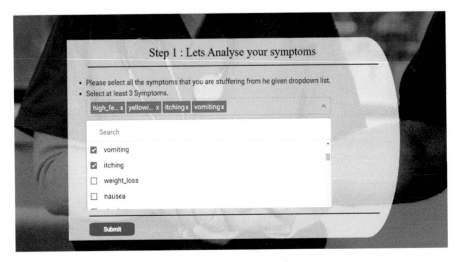

Figure 3.1   User interface to enter the symptoms.

```
},
"Thyroid": {
    "Symptoms": [
        "Fatigue",
        "Increased sensitivity to cold",
        "Constipation",
        "Dry skin",
        "Weight gain",
        "Puffy face",
        "Hoarseness",
        "Muscle weakness",
        "Elevated blood cholesterol level",
        "Muscle aches, tenderness and stiffness",
        "Pain, stiffness or swelling in your joints",
        "Heavier than normal or irregular menstrual periods",
        "Thinning hair",
        "Slowed heart rate",
        "Depression",
        "Impaired memory"
    ],
    "Test": {
        "name": "TSH Test",
        "format of upload": "PDF",
        "parameters": {
            "Total Triiodothyronine T3": {
                "type": "numeric",
                "sex range": false,
                "range": [
                    0.55,
                    1.65
                ],
                "low": {
```

Figure 3.2   JSON format with listed parameters (thyroid).

check for diseases left after filtering with the help of symptoms to verify that the entered symptom set is valid for that particular disease. If no matches are observed, the patient will be recommended home remedies for the entered symptoms. However, if symptoms of one or more diseases match, there is a possibility that the patient might be diagnosed with that disease. Predicting solely on symptoms is not reliable or accurate. Thus, our model suggests some tests that will confirm the suspicion of the disease.

b) Understanding test reports:

The patient needs to submit the suggested test report in the requested format (Figure 3.3). Test reports are of two formats:

i. Image:

For medical tests like CT scans, MRI scans, and X-rays, the system recommends that the patient upload the image of the scan. The convolutional neural network (CNN) is a class of deep neural networks usually used for classification. The proposed system uses CNN to classify images to decide whether the patient needs to consult a doctor or not [13]. The system generates a trained neural network model and weights for each disease. When a patient uploads the image of the scan of a particular test, the algorithm selects the weights of the trained model for that specific test. Before classifying the image, it is pre-processed using Keras and converted to the dimensions required by the specific model. The image is set as an input to the model weights, which classifies it and calculates a severity level. It helps to decide whether the patient should consult a doctor or not. Below is the detailed work of CNN on predicting pneumonia with a chest plain CT scan.

Pneumonia: For the detection of pneumonia, a chest plain CT scan is recommended; hence, we create a convolutional neural network trained on 5216 images. The neural network architecture has a total of 22 hidden layers, which includes 16 convolution layers, 5 max-pooling layers, and 1 flattening layer. After compiling all the layers, with the help of "Adam" optimizer, loss function is set to "binary cross-entropy." The model is trained with 15 epochs, with validation data of 614 images of both normal and pneumonia patients.

ii. PDF:

For medical tests like Bilirubin (Figure 3.4), N1 Antigen, RTPCR, etc., the system suggests that the patients upload the report in PDF format. "pdftotext" python library was used to extract information from a test report. The JSON file lists all the parameters to be fetched from the report (Figure 3.2). The text corpus might not precisely match the parameter name because of minor errors in converting PDF to text or poor image quality. The algorithm uses the Levenshtein distance

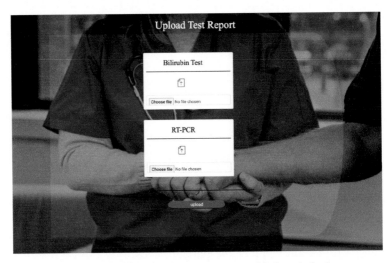

Figure 3.3   Uploading medical test reports with the required type.

Test		Result	Bio. Ref. Interval
Liver Function Test		**Sample Type**	**Serum**
Bilirubin-Total	:	0.62	0.1 – 1.2
Bilirubin-Direct	:	0.33	0 – 0.4
Bilirubin-Indirect	:	0.29	0.1 – 0.8
SGOT/AST	:	**41.16**	0.0 – 31.00
SGPT/ALT	:	**97.12**	0.00 – 31.00
Alkaline Phosphatase	:	246.26	80.00 – 290.00
Total Protein	:	6.01	6.0 – 8.0
Albumin	:	3.77	3.2 – 5.5
Globulin	:	**2.24**	3.2 – 5.5
Alb/GLB ratio	:	1.68	

Figure 3.4   Sample test format of PDF type (Bilirubin test).

ratio to search for unigrams and bigrams of the parameters in the text corpus [12]. After finding the match, the algorithm checks whether the parameter has a numeric or text value. Accordingly, a regular expression is used to locate the corresponding value for the parameter to make sure not to select any other numeric values mentioned in the report, like the range of a particular parameter that is written after the value. After extracting values, the data from the JSON file is used for generating a particular report based on the parameters that showed abnormal values. A final severity level is calculated for the specific test (Figure 3.5).

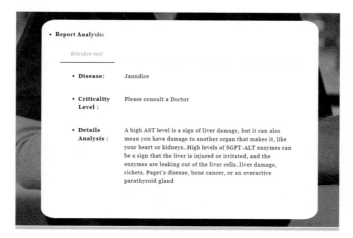

Figure 3.5 UI showing predicted disease, criticality level, and detailed analysis of the report.

3.2.2 Hospital monitoring system

A monitoring system for patients is necessary because it helps to identify early signs of patient deterioration, and timely treatment can increase the chances of saving patients' lives.

Generally, vital signs indicate the status of the patient's life-threatening functions in the ICU. The NICE guidelines recommend that heart rate (HR), respiratory rate (RR), blood pressure (BP), and oxygen saturation (SpO_2) are considered significant vital signs. Our proposed system monitors these critical signs, and the system is reliable, cost-effective, and easy to use.

Providing continual and flawless patient assistance is of maximal importance. We developed a technology for remote monitoring of ICU that consists of cameras and web-based applications [11]. High-definition cameras were installed over each ICU bed for monitoring systems round the clock and fetching images of the patient monitor in regular intervals (Figure 3.6). Once the system gets the fetched image, the extraction process is initiated. Computer vision algorithm was used for extraction of the required vital signs from the input image. It takes an image as input in either JPEG or JPG format. To identify all the text and numbers as string objects, it constructs a list of bounding boxes around all the recognized texts. Each element of the list, namely a bound, contains sub-elements such as accuracy with which the text gets classified, coordinates of the bounding box, and the recognized text itself. The required vital signs were compared with the standard values predicted by the algorithms.

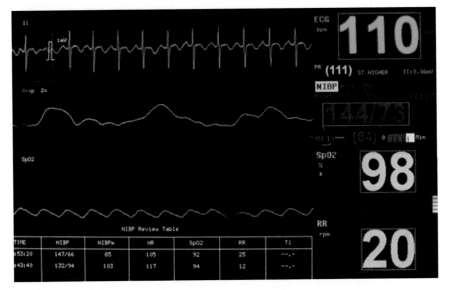

Figure 3.6 Patient monitor display while extracting vital signs.

Table 3.1 Standard vital sign range.

Vital signs	Range
Heart rate	60–100
Blood pressure	90–120/60–80
Respiratory rate	12–20
Oxygen saturation	95–100

The model considers multiple scenarios for setting the threshold values at the backend. Each procedure is explained in detail below.

3.2.2.1 The customized threshold for vital signs

A threshold value of vital signs varies from person to person on age, gender, and medical history. A person with no medical history will have standard vital signs as shown in the Table 3.1.

Threshold values for each vital sign are analyzed using different methodologies. Below is the list of all the vital signs with the detailed explanation of the algorithm for setting threshold values.

i. Heart rate (HR):

A fit heart rate differs from person to person. It depends on multiple factors like the age and physical fitness of the person. The heart rate gradually gets slower as a person moves through infancy toward senescence. The model uses random forest to predict the threshold value of

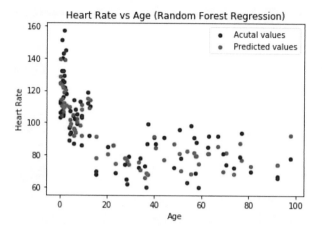

Figure 3.7 The graph represents the actual heart rate values of the patient and predicted values by the model.

HR depending on age (Figure 3.7). It creates 100 trees in a random forest, the model was further trained, and weights are adjusted according to the data values to get precise results.

ii. Respiratory rate (RR):

Age is a significant factor when it comes to an understanding of the respiratory rate. If the respiratory rate is below normal, it could indicate central nervous system dysfunction, or when above normal, it could designate another underlying condition. The model offers in-depth analysis of the random forest regression model, which predicts RR with respect to the age of the patient (Figure 3.8). It is pliable to fit the multi-range data without a need to specify the ranges manually.

iii. Blood pressure (BP):

Blood pressure is measured using two numbers. The first number was noted as systolic and the second as diastolic blood pressure. BP varies majorly with age and gender. According to the standard ranges, the model plots a graph for each age group. Further, a specific equation is generated for males and females individually for systolic and diastolic blood pressure that calculates the threshold values for each age. Following is the equation of systolic and diastolic blood pressure for males and females:

$$y = 1.62588x + 0.03015x^2 - 0.00402x^3 + 0.00009x^4 + 95.42763$$

$$(3.1)$$

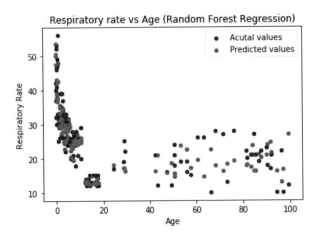

Figure 3.8 The graph represents the actual heart rate values of the patient and predicted values by the model.

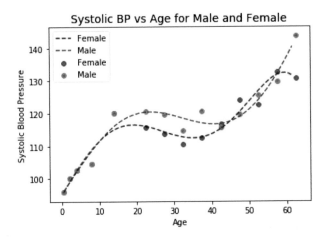

Figure 3.9 Systolic blood pressure and age for male and female.

$$y = 1.94579x + 0.00682x^2 - 0.00478x^3 + 0.00013x^4 + 94.95478,$$

$$(3.2)$$

where x represents the age of the patient and y represents the systolic pressure. Eqn (3.1) and (3.2) were used to calculate systolic blood pressure of males and females, respectively, which is visually represented in Figure 3.9.

$$y = 0.21755x + 0.12044x^2 - 0.00692x^3 + 0.00014x^4 + 59.57638$$

$$(3.3)$$

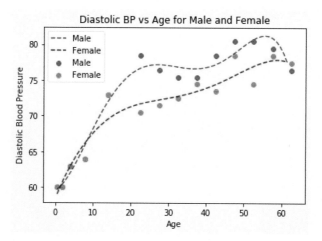

Figure 3.10 Diastolic blood pressure and age for male and female.

Table 3.2 BP range for blood-pressure-related disease.

Blood pressure	Range
Prehypertension	120–139/80–89
Hypertension	Above 140/90

$$y = 1.27654x - 0.03906x^2 + 0.00024x^3 + 0.00001x^4 + 58.39364,$$

$$(3.4)$$

where x represents the age of the patient and y represents the systolic pressure. Eqn (3.3) and (3.4) were used to calculate diastolic blood pressure of males and females, respectively, which is visually represented in Figure 3.10.

If the patient is suffering from a BP related disease the range of BP at rest is different than normal as shown is Table 3.2. The system allows the healthcare workers to update the standard values depending on the medical problems of the patient, like hypertension.

iv. Oxygen saturation (SpO_2):

A patient's blood oxygen level measures how well the body circulates oxygen from the lungs to the cells. SpO_2 is a reading that indicates what percentage of blood is saturated. The variation of regular reading and COPD or other lung diseases is shown in Table 3.3. The proposed model considers these variations and set the threshold values of specific patients accordingly.

Table 3.3 SpO$_2$ variation considering lung disease.

SpO$_2$	Range
Normal range	95–100
COPD or lung diseases	88–92

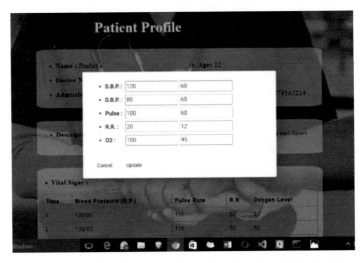

Figure 3.11 Patient profile page with an option to change the threshold value for a particular disease.

i. Vital signs of the patient may not be in the same range when they are suffering from any particular disease. For example, if a patient is suffering from severe kidney disease and blood pressure rises, that might indicate kidney failure. The model understands such scenarios for different diseases individually and sends alerts accordingly.

ii. Vital signs of a severely ill patient will improve as the patient recovers. Hence, it is necessary to update the threshold values by the condition of the patient (Figure 3.11). The designed system allows the doctors/nurses to change the threshold values, keeping the condition of the patient in mind.

Generally, patient monitoring information is available on the central station monitors, but if the central station is left unattended. Hence, the designed system sends an alert SMS on respective doctors' or nurses' smartphones (Figure 3.12) if an abnormality is observed by comparing extracted values from the patient monitor image and the threshold value set by the algorithm, the doctor, or the nurse.

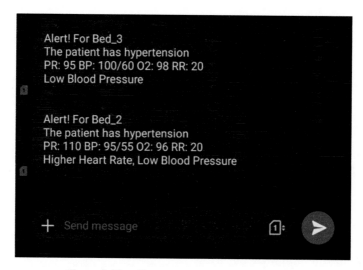

Figure 3.12   Alert message sent to hospital staff.

Table 3.4   Calculating the Levenshtein distance for five different reports and ratios.

Ratio	R1	R2
=1	3/3	3/9
≥0.95	3/3	5/9
≥0.90	3/3	5/9

3.3 Experimentation

3.3.1 The threshold for the Levenshtein distance

The algorithm looks for the parameter name in the text corpus. While extracting information from a PDF file, it is bound to have spelling errors as characters are misinterpreted. Thus, the Levenshtein ratio was used to calculate the edit distance between two words. Table 3.4 shows the ratio of the parameters that are correctly identified with respect to the Levenshtein distance ratio threshold.

3.3.2 The threshold for heart rate and respiratory rate

There are multiple algorithms used for prediction. According to the available data, the most appropriate models are random forest and decision trees. By training both the algorithms on the data, it was observed that there are more

Table 3.5   Comparing the errors of random forest regression and decision tree.

Algorithm	Mean absolute error (MAE)	Mean squared error (MSE)
Random forest	12.171	231.615
Decision tree	14.3	307.7

Table 3.6   Comparing the errors of random forest regression and decision tree for respiratory rate.

Algorithm	Mean absolute error (MAE)	Mean squared error (MSE)
Random forest	4.593	34.539
Decision tree	5.507	50.035

errors in the decision tree algorithm than in random forest, the results are tabulated in Table 3.5 and Table 3.6. Hence, the proposed model uses the random forest to minimize the errors.

3.4 Conclusion

The proposed system considers cost, ease of application, and accuracy [10]. The performance analysis shows that the proposed system is reliable and helpful due to its easy-to-use interface. This system is very beneficial during the current Covid-19 pandemic as it allows the patients to get diagnosed without visiting the hospital. The model gives better results as it analyzes symptoms and medical test reports. Analyzing medical test reports using CNN, optical character recognition, and natural language processing makes the model more flexible and accurate.

With the help of warning messages, sudden clinical deterioration was detected earlier, and on-site clinicians are alerted to intervene timely. With the outbreak of Covid-19, there is a scarcity of ICU beds, resources, and healthcare personnel. It is essential to monitor critical to monitor the critical patients carefully. The utilization of the proposed system will benefit the hospital staff and reduce the burden on healthcare workers.

References

[1] Sneha.R, Monisha S, Jahnavi C, Nandini S; "Disease prediction based on symptoms using classification algorithm" Journal of Xi'an University of Architecture & Technology, 2020

[2] Anshul Aggarwal, Sunita Garhwal, Ajay Kumar; "HEDEA: A Python Tool for Extracting and Analysing Semi-structured Information from Medical Records"Healthc Inform Res. April 2018

[3] Qing Li, Weidong Cai, Xiaogang Wang, Yun Zhou, David Dagan Feng, Mei Chen "Medical image classification with convolutional neural network" 2014 13th International Conference on Control Automation Robotics & Vision (ICARCV)

[4] Akash C. Jamgade, Prof. S. D. Zade; "Disease Prediction Using Machine Learning" International Research Journal of Engineering and Technology, May 2019

[5] Ahmed,Irfan, "Extract Valuable Data from PDFs With ReportMiner." Astera- Enabling Data-Driven Innovation, May 2021

[6] Benjamin Kommey, Seth Djanie Kotey, Daniel Opoku; "Patient Medical Emergency Alert System" International Journal of Applied Information Systems USA, December 2018

[7] Frank A. Drews; "Patient Monitors in Critical Care: Lessons for Improvement" Journal- Advances in Patient Safety: New Directions and Alternative Approaches

[8] B Naveen Naik, Rekha Gupta, Ajay Singh, Shiv Lal Soni & G D Puri "Real-Time Smart Patient Monitoring and Assessment Amid COVID-19 Pandemic – an Alternative Approach to Remote Monitoring" Journal of Medical Systems, 2020

[9] Kumar S, Merchant S, Reynolds R. "Tele-ICU: efficacy and cost-effectiveness of remotely managing critical care."" Perspect Health Inf Manag 2013

[10] Breslow MJ, Rosenfeld BA, Doerfler M, Burke G, Yates G, Stone DJ et al. "Effect of a multiple-site intensive care unit telemedicine program on clinical and economic outcomes: An alternative paradigm for intensivist staffing" Critcal Care Medicine, 2004

[11] Singh A, Naik BN, Soni SL, Puri GD. "Real-Time Remote Surveillance of Doffing during COVID-19 Pandemic: Enhancing Safety of Health Care Workers."Anesthesia & Analgesia: August 2020

[12] Rishin Haldar, Debajyoti Mukhopadhyay; "Levenshtein Distance Technique in Dictionary Lookup Methods: An Improved Approach"

[13] Nikhil Sonavane, Ambarish Moharil, Fagun Shadi, Mrunal Malekar, Sourabh Naik, Shashank Prasad; "Classification of Types of Automobile Fractures Using Convolutional Neural Networks" Machine Learning and Information Processing

4

Application of Machine Learning in Chest X-ray Images

V. Thamilarasi and R. Roselin

Department of Computer Science, Sri Sarada College for Women (Autonomous), India
Email: tamilomsiva@gmail.com; roselinjothi@gmail.com

Abstract

In early days, each and every field of medical image analysis and diagnosis took more time to conclude the medical report of patients. Most of the time, it resulted in failure, and it is a challenging task for the medical field. More than a decade's gap between technology and human society creates drastic loss in human life. In those days, radiologists were not able to make successful attempts in radiology. In any field, technology is the path for evaluation. In the early computer age, the recital level of AI and human intelligence started and tried to bind one another. In the initial stage, AI started with little success and now it is evidently surpassing the level of human intelligence. AI grounds its foot stronger to the widespread evaluation of medical imaging. In this technological era, the medical field uses a variety of techniques to examine every inner organ for diagnosis and treatment of the human body. Every medical testing needs a different type of image analysis in the form of X-rays, DICOM images, MRI scans, etc. All these images have different qualities and nature depending on patients and illumination around them. Medical image diagnosis and identification is a roughly tougher task for radiologists due to the different nature and modality of images, and it takes more time for analysis in different platforms and techniques. To prevent and cure diseases, physicians need prolonged support from radiologists and they need the hands of computer-aided diagnosis (CAD) tools and techniques. Machine learning is a branch of AI. Deep learning is a subset of machine learning and both areas are twins of AI. They are inseparable and, at the same time, each technology explores its medical analysis in different ways. This chapter explores application of machine learning and deep learning in chest X-ray image analysis. There is no limit in the use of machine learning techniques. It branches its root to maximum areas in the development of medical field. This chapter presents various machine learning techniques for segmentation and classification with case study.

4.1 Introduction

Medical image processing is a vast area. One cannot put any wall to define this as the limit for researchers to analyze. Everyone has to remember that this is the right situation to open the locks in the medical image analysis. According to physicians, modalities of image capturing are different for different disease diagnoses. As a result, it makes the radiologists more responsible to select correct and suitable techniques to process various images of patients. AI provides two ways that are considered a gift for radiologists: machine learning and deep learning. Each technology has its own space to analyze the object identification, detection, segmentation, classification, etc. Now radiology reaches its highest success in the timely analysis of medical images and reduces the death rate due to false and tedious result analysis. However, in medical images, chest X-ray images need special attention from radiologists. It is economic and is the most-used modality by physicians. Chest X-ray images are always a challenging domain in the medical field. In our body, most of the organs are present in our chest area and identification of every portion of the organs is critical, and it takes more time to separate a particular organ for analysis. Due to the different nature of patients and diseases, the illumination around them creates more impact in chest X-ray images.

Under the umbrella of machine learning, radiologists have to choose proper techniques for an accurate analysis and diagnosis. According to the position and the orientation, chest X-ray images have three views such as posterior–anterior, anterior–posterior, and lateral. Posterior–anterior (PA) is the most commonly preferred type and it is taken with the patient in a standing position to capture the images of heart and lungs. Anterior–posterior (AP) is less preferred because of its unclear nature. The lateral type provides a side view and it permits a three-dimensional analysis; it helps to identify the lesion. For analysis, researchers prefer anatomical order or mnemonic order. It is a wide area for radiologists to select a particular technology, which gives promising results for various analyses like airways, heart, lungs, diaphragm, bubbles, etc. In chest X-ray, all these are presented in white and black portions, and it is critical to identify them without suitable technology. Machine learning and deep learning jointly provide technological support for

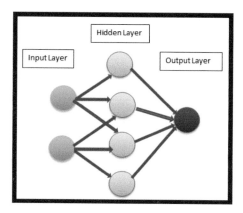

Figure 4.1 Simple structure of neural network.

radiologists and help physicians to recommend proper treatment for patients, thereby increasing the success rate of human life.

AI is the pillar that creates a strong base for medical image analysis. Neural network is the foremost technique that produces expected results in medical image analysis. Like the biological nervous system in human brain processing, well-connected neurons are present in artificial neural network. It works based on input, processing, and output. In the start of the digital era, ANN played a prominent role in various fields like engineering, computer science, object identification, wireframe modeling, and medical image analysis. Convolutional Neural Networks (CNN) follow a proper methodology for every image analysis, which includes pre-processing, image enhancement, edge detection, object identification, and localization, segmentation, and classification. It has a lot of techniques for exploring different images. It used stable algorithms for identification and problem solving among input, processing, and output. Feed-forward neural network, feedback networks, and self-organizing maps are different networks employed in ANN and learning happens in supervised or unsupervised methods. Figure 4.1 shows the structure of a simple neural network.

Feed-forward neural networks are always employed under supervised learning with back-propagation. Here, weights are altered based on bias, and it is gone repeatedly until expected output accuracy. It needs training and testing samples to proceed processing. Gradient descent and multi-layer perceptron are the most commonly used algorithms.

Feedback network works in both directions to get an equilibrium point, but it is a complicated one. Hopfield network is a special type of feedback network and is particularly used for tumor detection classification.

Self-organizing maps are a special type of ANN and its structure and algorithm are different from ANN. It is composed of a single-layer 2D grid of neurons and uses competitive learning to adjust its weights. The nodes do not know the weight of neighboring nodes. Kohonen neural network is of this type and it is particularly used for visualization.

CNN takes the basic concept of neurons from ANN and carries out miracles under machine learning and deep learning. The success of CNN branches under its structure, which has more layers, and it needs more labeled training data. So machine learning is a technology that helps to learn information directly from the given data. Deep learning is a specialized branch of machine learning. Machine learning plays a dominant role in solving critical problems in data analytics, big data, and medical image processing. This chapter elaborates the way that machine learning techniques perform analysis in chest X-ray images.

4.2 Chest X-ray Images

Chest X-ray images are a basic supportive visualization for physicians. Due to its economic nature and easy modality for handling, it has been preferred for most of the disease analysis. It is categorized into three types: posterior–anterior (PA), anterior–posterior (AP), and lateral. Figure 4.2 shows the types of chest X-ray images.

In chest X-ray (CXR), bones appear in white and other tissues appear in gray. CXR images help to identify the various pathological analyses including fracture detection, pneumonia identification, tumor analysis, Covid-19 identification, and other abnormalities in the chest area. Most often, CXR images are used for lung-disease analyses. There is a lot of datasets available for CXR images, which include Japanese Society of Radiological Technology (JSRT), NIH Clinical Centre dataset, Montgomery dataset, Shenzhen dataset, and pneumonia and Covid-19 dataset.

In 1998, JSRT and Japanese Radiological Society together produced the dataset with 247 images, which include 154 nodule and 93 non-nodule images for both lungs and the heart. It also contains left and right masks with ground truth images at a size of 2048×2048.

National Institutes of Health (NIH) produces datasets for clinical research. It includes 30,000 patients' CXR images.

Montgomery country dataset contains manually segmented lung masks that are produced by the Department of Health and Human Services in Montgomery. It contains 138 frontal CXR images with 80 normal and 58 abnormal images of tuberculosis.

PA	AP	Lateral

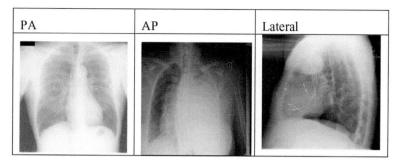

Figure 4.2 Types of chest X-ray images.

The Shenzhen dataset was developed by Shenzhen No.3 People's Hospital in China. It contains 662 frontal CXR images with 326 normal and 336 abnormal images of tuberculosis. It is in varying sizes and is provided in PNG format. Pneumonia and Covid-19 datasets are available in various repositories like kaggle, github, etc.

4.3 Literature Review

Yaniv Bar *et al.* (2015) proposed a deep learning method to inspect different pathological diseases and tested 433 images. This new model identified convolutional neural network with GIST features and gives improved performance over previous methods [1]. Jonathan Long *et al.* (2015) experimented with the fully convolutional neural network and different classification networks for segmentation [2]. Pauline *et al.* (2016) recommend an adversarial training method for training of semantic segmentation models. Convolutional semantic segmentation used to detect and correct the contradictions between ground segmentation maps and the result of segmentation nets and declare adversarial training approaches achieve better results for PASCAL VOC 2012 datasets [3]. Agus *et al.* (2017) proposed a new method for *k*-nearest neighbor and support vector machine by using Chan-Vese method and comparing results with old methods and finalizing these methods to provide improved accuracy with less sensitivity [4]. Alex *et al.* (2017) experimented with 1000 different classes of 1.2 million high resolution images in deep convolutional neural networks by ImageNet. The neural network used 60 million parameters with 650,000 neurons and five convolutional layers with max-pooling layer and Softmax function and dropout used to reduce overfitting [5]. Wora Wate *et al.* (2018) experimented with transfer learning for the classification of chest X-ray images for lungs. The lung cancer dataset used for analysis and the proposed method deliver 74.43% ± 6.01% for mean accuracy, 74.96%

± 9.85% for mean specificity, and 74.68% ± 15.33% for mean sensitivity. To identify lung nodule heat map used to identify the location, this method achieves expected results for a small dataset [6]. Hoel *et al.* (2019) presented the differentiable penalty, and it does not produce optimal solution, but it produces a better result than Lagrangian-based constrained CNN. This method focused on image size and image tags [7]. Pasa *et al.* (2019) proposed a new CNN method for tuberculosis screening in chest X-ray images of the Montgomery and Shenzhen datasets. The output of this CNN is analyzed with a saliency map and it is checked by a radiologist. Many cases find tuberculosis with minimum computational, memory, and power requirement [8]. Mohammad *et al.* (2020) used a deep learning model with data augmentation to test chest X-Ray images of Guangzhou women and children's medical center for pneumonia diagnosis. The model was evaluated based on both test accuracy and AUC score. This model achieves test accuracy as 98.435 and AUC score as 99.76 and results in a quick diagnosis of pneumonia [9]. Feng *et al.* (2020) experimented with optimized convolutional neural network for segmentation of medical images and adaptive distribution function used to solve the generalizability problems. Ultrasonic tomographic and lumbar CT medical images were used for this segmentation [10]. Arun *et al.* (2020) used artificial-intelligence-based classification for Covid-19 chest X-ray images. The size of the dataset is increased using 25 types of augmentation and the classification model is built by using a transfer learning method. Covid-19, non-Covid-19, pneumonia, and tuberculosis images for experiment and the declared proposed method attain more efficiency than the old method [11]. Hsin *et al.* explored convolutional neural networks (CNNs) with adaptive pre-processing for segmenting the lung region from chest X-ray images. Contrast enhancement, adaptive image binarization, and CNN were used for segmentation. This model achieves high training and reduces the image storage space [12]. Boran *et al.* (2020) experimented with 1583 healthy, 4292 pneumonia, and 225 Covid-19-confirmed images, and both machine learning and transfer learning models were used for experiments. Eightfold cross-validation is used for evaluation and 93.84% mean sensitivity, 99.18% mean specificity, and 98.50% mean accuracy have been achieved [13]. Dingding *et al.* (2020) proposed a new method with a combination of deep learning features and machine learning classification. These experiments were carried out by using Xception and SVM models with 1102 chest X-ray images and achieved 99.38% accuracy in SVM classifier which is better than the baseline Xception model, with an improved accuracy of 2.58%. Totally, the diagnostic accuracy was 96.75% and the authors recommend this system for Covid-19 analysis [14]. Lal *et al.* (2020) developed an AI imaging analysis tool to classify Covid-19 infection ($N = 130$), bacterial pneumonia ($n = 145$), and non-Covid-19 ($n = 145$) viral pneumonia chest X-ray images. Five supervised machine learning AI algorithms are used to classify and also two-class and multi-class are performed. Receiver operating characteristic (ROC) analysis is used for the classification model. This model achieved sensitivity, specificity, and accuracy of Covid-19 vs. normal as 100%, 100%, and 100%, bacterial pneumonia vs. Covid-19 as 96.34%, 95.35%, and 97.44%, and Covid-19 vs. non-Covid-19 viral pneumonia as 97.56%, 97.44%, and 97.67% [15]. Rajasenbagam *et al.* (2021) proposed a deep convolutional neural network to detect pneumonia infection in the 12,000 images of infected and non-infected chest X-ray images. Image annotation with the help of content-based image retrieval and data augmentation techniques was utilized by basic manipulation and deep convolutional generative adversarial network to increase the size of the dataset. Alex Net, VGG16Net, and deep CNN models were used to explore the experiment [16]. Jawad *et al.* (2021) used logistic regression and convolutional neural networks classifiers for fast and efficient results. Dimensionality reduction with principal component analysis was used to increase the learning process and classification accuracy. The generative adversarial network (GAN) is used for augmentation and this model achieves 95.2%–97.65% accuracy without PCA and 97.6%–100% with PCA for 500 Covid-19 images [17]. Adil *et al.* (2021) used optimal deep neural network and linear discriminate analysis for lung CT image classification and the classification attains accuracy of 94.56%, sensitivity of 96.2%, and specificity of 94.2% [20].

4.4 Application of Machine Learning in Chest X-ray Images

Machine learning (ML) is a branch of AI and teaches machines to learn like humans by learning from past experiences. It includes evaluating data, identifying patterns, and proceeding to solve the complex problems. ML used two types of techniques, supervised learning and unsupervised learning, to achieve the result. AI, machine learning, and deep learning afford more platforms for medical image analysis. Medical image needs more attention for image acquisition, image pre-processing, image enhancement, changing dynamic range of images, contour detection, restoration and smoothing of images, making 2D to 3D images, removing artifacts from the images, feature extraction, segmentation, classification, and object identification, detection, and visualization. AI provides a gateway to radiology to become a more promising area for multidisciplinary clinical diagnosis. It helps doctors to make keyhole surgeries without opening too much of the body. It helps

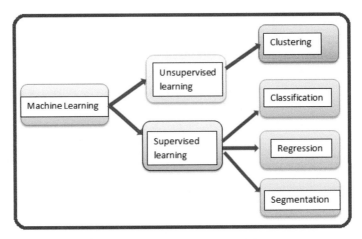

Figure 4.3   Machine learning techniques.

in interpretation of sensory information, web search analysis, self-driving automobiles, etc. In medical science, image processing delivers accurate and qualitative visualization of numerous medical image modalities like MRI, CT, X-ray, etc. Figure 4.3 shows the basic type of machine learning techniques.

Supervised learning used a collection of labeled data for input, and based on that input, it produces output.

Unsupervised learning is used to find the unknown pattern in the data. It learns experience and pattern from unlabeled data. It inherits the internal structure of unknown data. Clustering and dimensionality reduction belong to this category.

4.4.1 Clustering

Clustering finds hidden patterns in the data. It works like classification and does not need any previous data. It groups the similar data structure and then segments the particular portion. K-means clustering and fuzzy C-means (FCM) algorithms are the most commonly used method for grouping centroids. Hierarchical clustering, self-organizing maps, subtractive clustering, hidden Markov model, and Gaussian mixture models are some other clustering techniques.

K-means clustering partitions the image into k sections based on each section mean. The given data is divided into k clusters and finds the mean of every cluster. Based on the Euclidean distance, each data holds in a cluster. Finally, the k value is found for segmentation based on input and output

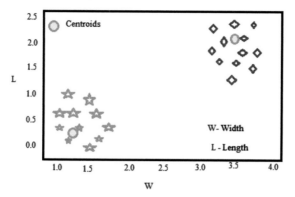

Figure 4.4   *k*-Means clustering.

k vectors. A drawback of this method is, it produces more clusters with noises, and initial valued need to be fixed. Figure 4.4 shows the model of *k*-means clustering.

FCM is most popularly used in medical image segmentations. It groups similar data values into the same group and also works based on the mean of each cluster. For corrupted images also, it provides better results.

4.4.2 Regression

Regression allows to predict continuous outcome variables based on predictor variables. Linear regression is the most commonly employed method for calculating continuous variables. It is denoted by

$$y = b \times x + c,$$

where *y* is the dependent variable, *x* is the independent variable, and *b* is the slope of best fit.

The output received by using more than one independent variable is called multiple linear regression. Decision tree, SVM, random forest, GPR, GLM, *k*-NN, ensemble methods, and neural networks belong to regression. Figure 4.5 shows the linear regression example. Linear regression is a relationship between a dependent (*y*) variable and one or more independent (*y*) variables. It predicts how dependent variables change according to independent variables. It is represented by a sloped straight line. It has two types: simple linear regression and multiple linear regression.

The decision tree as shown in Figure 4.6 composed of decision nodes with attributes that split the dataset, branches that specify the condition, and leaf nodes that represent the decision. The drawback of this method is its unstable algorithm.

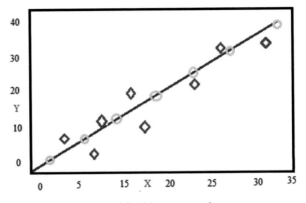

Figure 4.5   Linear regression.

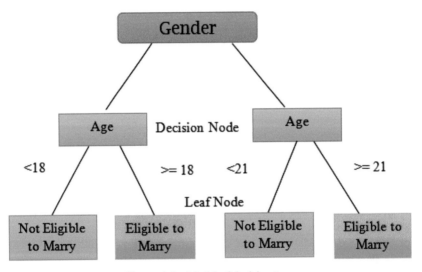

Figure 4.6   Model of decision tree.

Support vector machine (SVM) works well for classification and regression. It also performs both linear and non-linear regressions. It separates data into classes based on the hyper-plane. It finds the point closer to the hyper-plane and these are called support vector points. The calculation of distance between hyper-plane and support vector is called margin. Maximizing the margin is the goal of SVM. The optimal hyper-plane is derived from the maximum margin of the hyper-plane. Figure 4.7 shows the model of SVM.

Random forest works on the concept of multiple decision trees. First, *k* random points are selected, and the decision tree is built based on these data points. Second, the number of trees to build is selected and is repeated

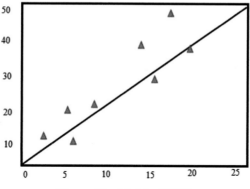

Figure 4.7 Model of SVM.

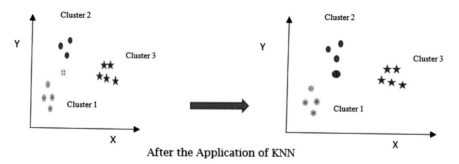

After the Application of KNN

Figure 4.8 Model of *k*-NN.

again. Finally, based on the input value, new data points are created for each tree-dependent variable and the average value of the predicted values is put to the final output. A drawback of this algorithm is that it needs more time to build trees.

The *k*-nearest neighbor model finds *k* most similar instances in the training set. The median value of the neighbor is used for input. A drawback of this model is that it takes more space for training the large data and also results in poor quality. Figure 4.8 shows the basic model of *k*-NN.

4.4.3 Segmentation

Segmentation is a core process of any medical image segmentation. Segmentation gives the exact location and extraction of a particular object. Cluttering is a widely used technique. Segmentation is magic and it changes the image information. Under the umbrella of CNN, segmentation is carried out successfully in machine learning.

Segmentation is a very important stage for any medical image segmentation. It is a process of annotation to group image pixels into a single group. It works the same as object detection, and segmentation results are based on the shape of the image that belongs to granular information. It simply divides images into regions of the same properties and it helps to distinguish objects from one another. It can be used in self-driving car detection, face detection, medical image segmentation, defect detection, etc. It is classified into two types: semantic segmentation and instance segmentation.

Semantic segmentation is a procedure of segmenting similar pixels into one group. Instance segmentation is used to segment if there is more than one object in the image. Here, the detected object is masked with a color and pixel related to those objects specified with the same color. So different objects are denoted in different colors.

Based on segmentation, only physicians decide the nature of treatment recommended for the patients. Depending on the type of operation, the segmentation performs experiments and produces high-accuracy results. CNN with encoder and decoder structure is the reason for success of segmentation in machine learning.

4.4.4 Classification

Classification is a specialized approach to assigning pixels in the image into a particular category. The data and classes play a significant role in classification. It belongs to supervised learning. It works based on learning techniques and feature selection sets. It is a method of recognizing, understanding, and grouping the objects based on categories. It is one of the pattern recognition techniques used to find patterns in given data. Based on given labeled data, it classifies the malignancy. Support vector machine, linear discriminant analysis, naïve Bayes, k-NN, decision tree, and neural networks are some of the classification methods in machine learning.

k-Nearest neighbor (k-NN) works based on k-nearest neighboring data points. The k value plays a major role in improving accuracy.

Support vector machine (SVM) is one of the supervised learning algorithms; it can be used for both regression and classification. This algorithm features data point plots in n-dimensional space and classifies to identify hyper-planes that differentiate the two classes. Coordinates are called support vectors and SVM plays an eminent role in classification. SVM creates a decision boundary in n-dimensional space; it leads us to put data in the correct category. Figure 4.9 shows the simple classification of nodule and non-nodule lung chest X-ray images.

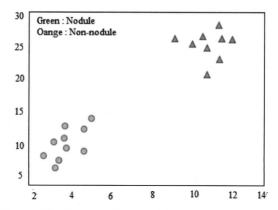

Figure 4.9   Nodule and non-nodule classification of SVM.

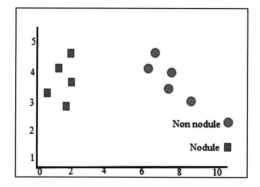

Figure 4.10   Nodule and non-nodule classification by naïve Bayes.

Linear discriminant analysis used the dimensionality reduction concept for classification. It is a pre-processing step for pattern classification. It minimizes the overfitting and computational cost. Once the dependent variable is selected, then the related feature information dragged out from the existing dataset to check the dependent variable. By this way, dimensions are reduced and special features are retained in the new dataset.

Naïve Bayes classifiers are simply a collection of classification algorithms that work based on the Bayes theorem. Feature matrix and response vectors are two parts of naïve Bayes classification. Feature matrix contains the dependent features. Response vector contains the value of a class variable. So it calculates the chance of a given data point belonging to some category or not. Figure 4.10 shows the model of the classification of nodule (1) and non-nodule (0) lung X-ray images in scatter plot.

Decision tree has the ability to order classes at a precise level. It resembles a flowchart and creates categories within the categories. It simply

Table 4.1   Accuracy of decision tree classifiers.

Classifiers	Accuracy
Random forest	82.59%
Simple cart	80.97%
Random tree	79.75%

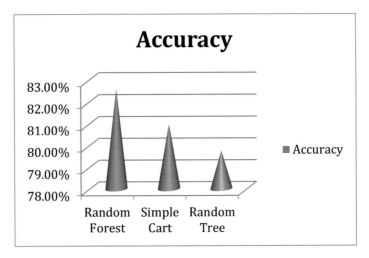

Figure 4.11   Accuracy from decision tree classifiers.

represents human thinking. It can be easily understood due to its tree-like structure. It has three nodes: root node, decision node, and leaf node. Many decision tree classifiers like random tree, random forest, and simple cart perform classification of lung CXR images and JSRT dataset taken for analysis. Table 4.1 shows that random forest produces higher accuracy than other methods, and Figure 4.11 shows the graphical representation.

Random forest algorithms are the development of decision trees; here also, decision trees are constructed for training data and new data placed in any of the trees as random forest. It solves the problem that is not solved by the decision tree. It works very well in medical data. It can handle large sets of data and prevent the problem of over-fitting.

A neural network simply resembles human brain activation. ANN contains neurons that are responsible for layers. Neurons are called tuned parameters and the output layer is called terminal neurons. It contains more layers and each layer output is transferred to the next layer. Each layer has different non-linear activation functions.

4.5 Case Study: Lung Chest X-ray Images

This case study explores the role of machine learning in lung chest X-ray images. In radiology, lung chest X-ray images need more help from the computer-aided diagnosis and techniques. In today's world, lung cancer nearly kills people of all age groups and it increases day by day due to unhealthy habits and pollution. There are a lot of segmentation techniques available for medical image analysis. Every technique has its own space. It is not suitable for all types of images. Segmentation is the root of all other analyses in image. Proper selection of segmentation techniques may help physicians do their further procedures. Accurate detection of lung portion leads to exact finding of diseases present in lung portion. A proper segmentation is only able to produce the exact location and object identification in lungs. Better prediction leads to reducing the range of failure rate in disease diagnosis, thereby reducing the death rate of human society. In recent years, machine learning techniques spread their magic in maximum fields. Particularly, the helping hands of machine learning techniques prominently reduce the failure. Machine learning techniques produce results within a time limit and yield expected accurate results.

4.5.1 Methodology

The methodology involves the following steps in segmentation and classification of lung images. For every medical analysis, some steps are compulsory to follow in nature. Pathological tests need more precise images to identify the fault areas in the human body. Everything in medical image analysis selects and follows timely procedures; then only, they effectively treat the patients in the correct way. Every field arises to save people in some way to live their life peacefully. Physicians and radiologists trust the methods used in the machine learning techniques in which they provide timely help for the human world. Machine learning techniques reach their effectiveness through CNN with different activation functions, number of filters, Softmax function, etc.

Digital images are processed with the help of computer algorithms through image processing. It has a wider range of algorithms to process images to enhance the image quality and important features.

An image is composed of a two-dimensional array in the range of 0–255. It can be represented as $f(x, y)$ in terms of mathematics. x and y are horizontal and vertical coordinates, which provide pixel value of image in

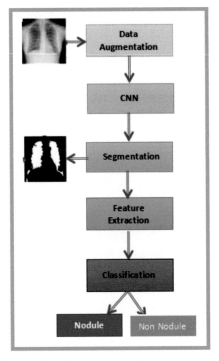

Figure 4.12   Methodology of machine learning.

that point. It involves some basic steps to proceed with problem solving. Figure 4.12 shows the methodology of machine learning.

1.   JSRT dataset

2.   Image pre-processing

3.   CNN

4.   Segmentation

5.   Feature extraction

4.5.2 JSRT dataset

In 1998, the Japanese Society of Radiological Technology (JSRT) and the Japanese Radiological Society (JRS) combined and developed the dataset with and without chest lung nodules. Most of the research scholars used this dataset for their analysis worldwide. The dataset contains 154 images of nodules and 93 images of non-nodules; totally, it has 247 images at a size of 2048 × 2048. But machine learning techniques deliver the best results for small-size images. Data augmentation has been done as a pre-processing step.

JSRT also provides an ImageJ tool that provides more options for image scaling and rotation operation. First, the image was received from the dataset by using this tool; for that, we have to select the raw image and save it in .jpeg or .png format, whichever format that needs to be processed. There are a lot of other options available. First, the given image is converted into a gray-scale image, the size is reduced to 512 × 512, and horizontal and vertical flipping and rotation are done by 45°, 90°, 180°, and 360° by the ImageJ tool.

Both nodule and non-nodule images are read from the dataset by using the ImageJ tool. Lung chest X-ray images need careful handling in pre-processing step to apply machine learning techniques.

4.5.3 Image pre-processing

Read the image: Before going through the image, we have to read the image for processing. Check whether all images are in the same format and required format. Many processes need only gray images rather than color images, which allow easy recognition of objects. The color images contain more than gray images. Some operation needs color images; that time, we have to convert gray to color images.

Resize the image: Most of the datasets do not contain images in the same size. ML algorithms need scaling options to resize the images. Small-size images work better than big dimension images. Particularly, CNN produces best and quicker results for unified dimension dataset.

Remove the noise: Machine learning algorithms do not need so much noise-removal procedures. But some images need background only, and some need foreground only; for that purpose, noise removal is used to make some adjustments in the image.

Data augmentation: The accuracy of machine learning techniques lies on amount data being given as a training set. Rotation, scaling, and transformation are some data augmentation techniques being used to increase the size of the dataset. Rotation in different degrees, various scaling, and transformations are used to increase the number of images in the dataset. The process of creating new data from given available data is called data augmentation. Augmentation helps in various ways to improve the result of machine learning segmentation and classification. Figure 4.13 shows the data augmentation in a lung CXR image.

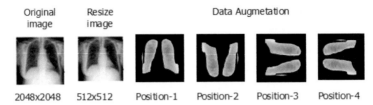

Original image	Resize image	Data Augmetation			
2048x2048	512x512	Position-1	Position-2	Position-3	Position-4

Figure 4.13 Pre-processing in a lung CXR image.

4.5.4 CNN

Convolutional neural network (CNN) is a basic architecture for machine learning techniques. The main goal of CNN is to retain important features in the images. Machine learning techniques always need more data to process; then only it can produce accurate results. In the medical field, every data is different and most of the dataset is not enough to handle machine learning techniques. Data augmentation has been done by using various pre-processing techniques.

CNN used the same padding to increase the dimensionality and valid padding to decrease the dimensionality. Max-pooling is used to down-sample the input representation, which reduces the learning time and reduces computational cost. It selects important features by using filters and reduces the number of pixels for output and shortly reduces dimensionality. Average pooling calculates the average of each patch of the feature map. It takes a given image feature map as a square and it smoothens the image without considering the important features of the image. Hence it down samples the feature map.

Stride represents the number of filters moved in pixels at the input screen. If it is 1, it moves the filters by one pixel at a time, and if it is 2, it moves the filters by two pixels at a time. On some occasions, the filters do not properly work; so time padding is used. Zero padding and average padding are two types of time padding.

Padding helps to change the input images to fit our requirements. In zero padding, it simply adds 0 in edges. It avoids the shrinking of the image after filtering. In the same padding, there is no change in the input and the output image.

Batch normalization normalizes the input and stabilizes the learning process. It reduces the number of epochs required to train the networks.

Activation functions help networks to learn complex data. It decides whether the neurons are activated or not. Following are some of the activation functions:

- Binary step function

- Sigmoid function

- Softmax

- Tanh

- ReLU, etc.

ReLU is the most commonly used basic activation function. It allows the model to perform quicker and produce a better result. It is mostly preferred by CNN. For negative output, it returns 0 and for positive output, it returns the value as x. At value 0, it becomes non-linear or biased. It is also called interactions or non-linearities.

$$\text{ReLU equation} = A(x) = \max(0, x). \qquad (4.1)$$

Softmax changes the vector of k real values into vectors of k real values that sum to 1. It is included in the final layer of the network.

Fully connected layers help to learn high-level features from the output of convolution layers. These are some basic needs to work CNN. Machine learning architecture is based on these CNN structures.

Lung chest X-ray images provide better results for semantic segmentation. Convolutional layer with same padding, stride 1, max-pooling, and max-pooling with ReLU activation function work well for lung CXR images. The accuracy is nearly 82.90%.

UNET-based architecture also provides better results for segmentation and classification. UNET architecture contains up-sampling and down-sampling paths with 3×3 convolution layer with max-pooling, same padding, stride 1, and ReLU activation function. After segmentation, the features are extracted and applied for classification. Dice coefficient is used as evaluation metric for both segmentation and classification. It achieves nearly 89% accuracy in segmentation [19] and 98% in classification [18].

Figure 4.14 shows the semantic segmentation of lung chest X-ray images.

4.6 Conclusion

Medical images are the backbone of radiological analysis. Timely detection of any disease can easily cure it. Many times, it is a challenging task for radiologists and sometimes it results in failure. Physicians need extra support from radiologists and radiologists need help from computer-aided analysis and diagnosis. Analysis of chest X-ray images depends on the technique being adopted. Machine learning techniques are a gift for the medical world.

Grey image   Segmented   Ground
                 lung   truth

Figure 4.14   Semantic segmentation
                  of lung CXR image.

It increases the lifespan of every human being in the society. Machine learning algorithms are based on pre-defined featured algorithms with explicit parameters and expert knowledge. Statistical machine learning models like support vector machine and random forest are also used for this analysis. Advancement in CAD techniques includes neural network which is the base for AI and convolutional neural network is the bridge between AI, machine learning, and deep learning. Artificial intelligence (AI) health innovation applications are an umbrella for the medical field. Unalterable increase in the amount of medical data, which is not easily identified by the human eye, needs more support from AI. Radiologists are the basic pillar of this digital medical environment. Nowadays, radiology moves to objective science rather than depending on subjective perceptual skills. Accurate analysis of such images helped in saving human lives. To accomplish this, AI established its roots in machine learning and deep learning. The huge volume of data in medical image analysis utilizes tools of AI to recover hidden points, affected areas, fractures, cardiovascular problems, eye problems, and cancer detection in every part of the body. Machine learning techniques are preferred in medical image analysis because it does not need much pre-processing and does not take much time. Segmentation is the basic mechanism for image processing, and based on that only, other process have produced good results. Proper segmentation leads to feature extraction for classification. Classification promotes physicians to identify nodule and non-nodule and cancerous and non-cancerous parts in lung chest X-ray images.

4.7 Future Study

Machine learning methods work well for most of the medical images. For lung images, only very few architecture were experimented to produce the expected result. Lung CXR images need more experiment by various architecture in future. Machine learning techniques work well in large datasets, but for lung CXR images, there is no such standard dataset; so this experiment needs a large volume of datasets.

References

[1] Yaniv Bar, IditDiamant, LiorWolf, Sivan Lieberman, Eli Konen, Hayit Greenspan, "Chest pathology detection using deep learning with non-medical training". In: IEEE International Symposium on Biomedical Imaging (ISBI).2015.

[2] Jonathan Long_ Evan Shelhamer_ Trevor Darrell, "Fully Convolutional Networks for Semantic Segmentation", IEEE Conference on Computer Vision and Pattern Recognition, pp.3431–3440, 2015.

[3] Pauline Luc, Camille Couprie, Soumith Chintala," Semantic Segmentation using Adversarial Networks", arXiv:1611.08408v1 [cs. CV] 25 Nov 2016.

[4] Agus Pratondo, Chee-Kong Chui, Sim-Heng Ong, "Integrating machine learning with region-based active contour models in medical image segmentation", journal of visual communication and image representation, Volume 43, Issue: C, pp. 1–9, 2017.

[5] Alex Krizhevsky, Ilya Sutskever, Geoffrey E. Hinton, "ImageNet Classification with Deep Convolutional Neural Networks", Communications of the ACM, Volume.60, No.6, pp. 1–9, 2017.

[6] Worawate Ausawalaithong, Arjaree Thirach, Sanparith Marukatat, Theerawit Wilaiprasitporn, "Automatic Lung Cancer Prediction from Chest X-ray Images Using the Deep Learning Approach", The 2018 Biomedical Engineering International Conference (BMEiCON-2018).

[7] Hoel Kervadec, Jose Dolz, Meng Tang, Eric Granger, Yuri Boykov, Ismail Ben Ayed, "Constrained-CNN losses for weakly supervised segmentation", Medical image analysis, Volume 54, pp. 88–99, 2019.

[8] F. Pasa, V.Golkov, F. Pfeifer, D. Cremers & D. Pfeifer, "Effcient Deep Network Architectures for Fast Chest X-Ray Tuberculosis Screening and Visualization",Nature Research (Springer Nature), 9:6268, 2019.

[9] Mohammad Farukh Hashmi, Satyarth Katiyar, Avinash G Keskar, Neeraj Dhanraj Bokde and Zong Woo Geem," Efficient Pneumonia Detection in Chest Xray Images Using Deep Transfer Learning", Diagnostics, Volume 10(6): 417, pp. 1–23, 2020.

[10] Feng-Ping An and Jun-e Liu, "Medical Image Segmentation Algorithm Based on Optimized Convolutional Neural Network-Adaptive Dropout Depth Calculation", Hindawi Complexity Volume 2020(7), Article ID 1645479, pp. 1–13,2020.

[11] Arun Sharma, Sheeba Rani, and Dinesh Gupta, "Artificial Intelligence-Based Classification of Chest X-Ray Images into COVID-19 and Other

Infectious Diseases", Hindawi International Journal of Biomedical Imaging Volume 10, pp. 1–10, 2020, Article ID 8889023.

[12] Hsin-Jui Chen, Shanq-Jang Ruan, Sha-Wo Huang and Yan-Tsung Peng, "Lung X-ray Segmentation using Deep Convolutional Neural Networks on Contrast-enhanced Binarized Images", Mathematics, 8, 545, p. 1–13, 2020

[13] Boran Sekeroglu1 and Ilker Ozsahin, "Detection of COVID-19 from Chest X-Ray Images Using Convolutional Neural Networks", SLAS Technology, Volume. 25(6), pp. 553–565, 2020.

[14] Dingding Wang, Jiaqing Mo, Gang Zhou, Liang Xu, Yajun Liu, " An efficient mixture of deep and machine learning models for COVID-19 diagnosis in chest X-ray images", PLOS ONE journal.pone.0242535. pp. 1–15, 2020.

[15] Lal Hussain, Tony Nguyen, Haifang Li, Adeel A. Abbasi, Kashif J. Lone, ZirunZhao, Mahnoor Zaib, Anne Chen and Tim Q. Duong, "Machinelearning classifcation of texture features of portable chest Xray accurately classifes COVID19 lung infection", Hussain et al. BioMed Eng OnLine, 19, 88, pp. 1–17, 2020.

[16] T. Rajasenbagam, S. Jeyanthi, J. Arun Pandian, "Detection of pneumonia infection in lungs from chest Xray images using deep convolutional neural network and contentbased image retrieval techniques", Journal of Ambient Intelligence and Humanized Computing (Springer), 2021.

[17] Jawad Rasheed1, Alaa Ali Hameed1, Chawki Djeddi, Akhtar Jamil, Fadi AlTurjman, "A machine learningbased framework for diagnosis of COVID19 from chest Xray images", Interdisciplinary Sciences: Computational Life Sciences, 13, pp. 103–117, 2021.

[18] Thamilarasi, V. Roselin. R, " Automatic Classification and Accuracy by Deep Learning Using CNN Methods in Lung Chest X-Ray Image", IOP Conf. Ser.: Mater. Sci. Eng. 1055 012099, 2020.

[19] Thamilarasi, V. Roselin. R," U-Net: Convolution Neural Network for Lung Image Segmentation and Classification in Chest X-Ray images", INFOCOMP, Volume. 20, no. 1, pp. 101–108, June, 2021.

[20] Adil Khadidos, Alaa O. Khadisos, Srihari Kannan, Yuvaraj Natarajan, Sachi Nandan Mohanty, & George Tsaramirsis, "Analysis of COVID-19 Infections on a CT Images Using DeepSence Model", Frontiers in Public Health, (2020). Doi : doi.org/10.3389/fpubh.2020.599550.

5

Integrated Solution for Chest X-ray Image Classification

Nam Anh Dao[1], Manh Hung Le[1], and Anh Ngoc Le[2]*

[1]Electric Power University, Vietnam
[2]Swinburne Vietnam, FPT University, Vietnam
Email: anhdn@epu.edu.vn; hunglm@epu.edu.vn;
*Corresponding author's email: ngocla2@fe.edu.vn

Abstract

In this chapter, we present a join solution capable of identifying the symptoms of Covid-19 in lung X-ray images. The design is to study the components of an automatic medical imaging system suited for detecting signs of defection by Covid-19 in the lung area of the images. In contrast to the conventional implementation, the proposed method allows analytical convolutional neural networks (CNNs) to be integrated smoothly with other learning methods. It relies on feature extraction from the medical images to represent latent features as a result of convolution's implication. These features are proposed to be discovered by CNN, which allow the potential detection of abnormal image patterns. We found that within our case study, feature reduction is being adopted for the development of a new efficient realization for a join solution of CNN and support vector machines (SVMs). We demonstrate that features derived from CNN, its dimensions reduction, and classification are comprehensively explainable by Bayesian inference to capture the fundamental analysis flow enabling the classification of medical images.

5.1 Introduction

There has been increased interest in computer vision systems for medical imaging applications. In light of applying machine learning in the analysis of medical images, complex objects can be recognized programmatically by different methods. One of the most particular applications is to provide a diagnosis with medical image analysis, by that image objects are detected and then analyzed for their classification [1–3]. It has been noted that chest X-ray has been mentioned as a significant tool for clinical identification and computer-aided diagnosis has its high importance for supporting medicines in diagnosis and tracking disease history. In considering the high speed of the Covid-19 spreading particularly, vaccines are not enough for developing countries; the medical image analysis can provide effective tools for medical diagnosis and treatment. We address the problem of computer-aided diagnosis with the intention of detection of defection signs of Covid-19 in chest X-ray images.

The method presented in this chapter also attempts to perform an analysis of chest X-ray images for recognizing Covid-19 defection in the lung area of chest X-ray images, whilst providing an essential framework that allows for accessing latent features of neuron networks for further integrated learning solutions. The target is to enhance the reliability of the diagnosis by medical imaging.

In Section 5.2, we perform a survey of related work that provided solutions for the question of classification. Works with deep learning are especially interested in this study with creative improvement. We give a brief description of the proposed method by Bayesian framework where explicit explanation by learning inference flow is provided in Section 5.3. In contrast to many existing methods, however, our method copes efficiently with fundamental components of the learning process. This is achieved by extracting features that CNN can supply and by analyzing the features for their classification capability. We then implement a learning method to get the final classification. Experiments and results are reported in line with the described method in Section 5.4.

5.2 Related Work

We first discuss techniques that employ deep learning for the identification of symptoms or defection in the lung area from chest X-ray images (I). The availability of other learning methods for assistance of the deep learning is also addressed in this section (II).

(I) Zhao *et al.* [4] have created an open source database, which has more than 300 Covid-19 computed tomography (CT) images from 216 Covid-19 positive and 463 Covid-19 negative CTs. A diagnosis technique based on self-supervised learning and multi-task learning was proposed for diagnosing Covid-19 patients. An accuracy of 0.89 was reported by the study.

The hybrid deep learning model of Nandi *et al.* [5] represents a combination of ResNet50 and MobileNet. Here, the ResNet50 [6] is a light architecture of a deep convolutional neural network containing 50 layers to formulate learning residual functions that refer to the layer inputs. In [7], MobileNet is a proposed deep learning model of lightweight using depth sensible independent convolutions where a single convolution is performed on a color channel rather than integrating all channels and pressing it down as 2D CNN. The hybrid method combines identity of the depth-wise separable convolutions of MobileNet and mapping of ResNet50, which can diminish the vanishing gradient problem and can boost the gradient backward flow in the network.

To produce CT image features for diagnostic purposes, Wang *et al.* [8] implemented an inception migration neural network in a retrospective, multi-cohort, and diagnostic study. The method can particularly shorten the detection time for disease supervision and can reduce the diagnostic workload of physicians in the field. The accuracy of 82.9% was reported by the study and the accuracy for test with external samples is delivered by 73%.

In a study of dataset of chest X-ray images provided by the National Institutes of Health Clinical Centre, USA with labels contributed by an actual research from Google Health [9], Azemin *et al.* [10] have applied a deep learning model with the foundation of the ResNet-101 [11] that has convolutional neural network architecture. The network was trained to recognize objects from thousands of images and then retrained to discover dissimilarity in chest X-ray images. It has been shown that labels that are related with Covid-19 cases were used with the unique publicly accessible data for learning. As part of performance, area under the receiver operating curve (ROC), accuracy, specificity, and sensitivity were reported as 82%, 71.9%, 71.8%, and 77.3%, respectively.

Ahmed *et al.* [12] performed a study on an X-ray image database whose labels are viral pneumonia, Covid-19, and normal [13], [14], which were created in collaboration by hospitals and physicians [15]. They proposed a CNN model containing five convolutional layers. Each layer was followed by batch normalization, max-pooling layers, and dropout. Performance was reported with an *F*-score of 89.88% and an accuracy of 90.64% for a cross-validation by five folds.

We follow the CNN approach addressed by the above-mentioned works for Covid-19 classification of database of chest X-ray image, however with modifications for performance improvements. This is based on a twofold concept: selection and combination. We will apply feature selection for image features and implement a combination of CNN with SVM.

(II) The techniques used to achieve an optimized accuracy value: Hasmadi *et al.* [16] studied the influence of batch size and image size for web-based learning applications. Thus, smaller values of image size are proposed to be used for training in Google Colab avoiding crash. For such learning sessions, three batch sizes are used for comparison consisting of 64, 128, and 256.

To perform machine learning for Covid-19 symptoms from chest X-ray images, Tho *et al.* [17] showed a class balancing algorithm that used a detector of imbalanced samples to explore a minority class. Thus, images were trained and tested by VGG16 deep learning model to get initial image features. Then for such training data, the use of imbalancing methods was implemented to get class balancing. In the final step of this study, SVM was proposed to detect any sign of defect of corona in chest X-ray images.

Xi *et al.* [18] proposed a twofold sampling awareness network to predict Covid-19 signs from the Community Acquired Pneumonia in chest CT images. So, an awareness program with a 3D convolutional network was addressed to search in the infection regions of lungs. Here, imbalanced distribution of the sizes of the infection regions was found for distinction between Covid-19 and community acquired pneumonia due to the quick recognition of Covid-19 after the symptoms rise. To mitigate the imbalanced learning, the authors applied a dual-sampling strategy.

For a primary screening technique to discriminate Covid-19 pneumonia from healthy cases and Influenza-A viral pneumonia with CT images, a deep learning technique was applied by Xu *et al.* [19]. A 3D learning method from pulmonary CT image dataset was proposed to detect candidate infection regions. Then the image regions were categorized with scores into Influenza-A viral pneumonia, Covid-19, and irrelevant to infection groups by a classification method with focusing on location attention. This helped to estimate the infection type and full trust result of the image. It has been shown that the averaged accuracy achieved 86.7 % from all the studied CT image samples.

With the development of deep learning with the assistance of other learning methods, the above-reported works provided solutions for the question of Covid-19 diagnostics support by chest X-ray image classification. We also try to search a new solution by combining the ability of CNN for extraction of latent features and the possibility of effectively filtering the features to choose the features that respond the best to the classification. The method and its foundation of inference are described in detail in Section 5.3.

5.3 The Method

One of our primary concerns is in facilitating exploration and analysis of medical X-ray chest images when signs of defection by Covid-19 are threatening effective breathing and causing health loss. The main challenge is that there are distinctive types of lung diseases, and signs of defection are not always malicious and the good guy/bad guy binary classification does not precisely implement.

5.3.1 Feature extraction

We describe the techniques used to detect defects in each lung X-ray image marked by . Latent features to associate the image also can be analyzed with Bayesian method published by Barber [20], allowing to represent conditional probability for an image sample and class , which is a member of a set of classes covering positive by 1 or negative by 0

$$p(c|a) = \frac{p(a|c)p(c)}{p(a)}. \tag{5.1}$$

Given the rather 2D gray chest X-ray image a, its maximum *a posteriori* (MAP) for the most likely class is also determined by a Bayesian choice with the set of classes

$$c_{\text{MAP}} = \text{argmax}_{c \in C} \ p(c|a). \tag{5.2}$$

The most likely class c is associated with the image if Bayes' rule (5.1) allows to determine $p(c|a)$.

$$c_{\text{MAP}} = \text{argmax}_{c \in C} \frac{p(a|c)p(c)}{p(a)}. \tag{5.3}$$

However, the denominator $p(a)$ is usually suggested to be dropped

$$c_{\text{MAP}} = \text{arg max}_{c \in C} p(a|c)p(c). \tag{5.4}$$

The schema (Figure 5.1) shows a chest X-ray image a with patterns of lung region that can be defeated by diseases, while class indicates positive or negative.

Consider now a set of latent features f of a chest X-ray image f discovered by learning process, permitting associate features f to an image in a form of a vector:

$$f = \text{latent\_feature}(a), \ f \in \mathbf{R}^1. \tag{5.5}$$

Therefore, the joint probability distribution $p(c, a)$ can be seen in detail, where the probability of class c given features f and measure of feature f given image sample a are introduced:

$$p(c,a) \approx \sum_f p(c|f) p(f|a). \tag{5.6}$$

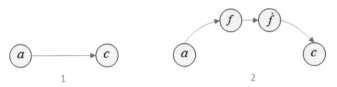

Figure 5.1   (1) Relation between chest X-ray image and class . (2). Image feature is determined by CNN on and then is reduced to have

In the following, a detailed information of an image region around each pixel x of the image a is needed for convolution operation with kernel h that has a size of $(2k + 1) * (2k + 1)$:

$$f(x) = h^*a(x) = \sum_{dx=-k}^{k} h(dx)\, a(x + dx).$$ (5.7)

As the operation of convolution is the foundation of convolutional neural networks that have multiple hidden layers, variable CNN models can be studied to implement for creating features f in eqn (5.7). Naturally, this helps to discover latent features from original chest X-ray images. The models are obviously computationally costing a lot. However, their deep learning at multiple convolution levels is produced and particular features for getting the measure of classification can be created. As will be seen, in our study, the task of feature creation (5.7) is carried out by convolutional neural networks including AlexNet and VGG19.

The convolutional neural network named AlexNet was designed by Alex Krizhevsky *et al.* [21] for a contesting of the ImageNet Large Scale Visual Recognition Challenge in 2012. A particular design with the depth assisted the model to yield a top-5 error of 15.3%, which was more than 10.8 percentage points lower than that of the runner-up in the contest. The high performance was supported by the utilization of graphic processing units (GPUs) during training for overcoming computational expensiveness.

The convolutional neural network design titled "VGG16" by Simonyan and Zisserman [22] was shown to give improvement over AlexNet by resetting large kernel-sized filters. The sizes in the first two convolutional layers were set to 11 and 5. There are multiple filters of 3 × 3 kernels. This model was trained for the ImageNet, which is a database having 1000 classes distributed for 14 million images. Thus, the test accuracy of 92.7% and top-5 place were achieved.

The high performance was supported by NVIDIA Titan Black GPU. Having defined a CNN that have 16 convolutional layers containing five pooling layers and three fully connected layers, VGG19 [23] proposed a deep learning design with 19 layers for the ImageNet.

5.3.2 Feature reduction

To make the classification more effective, we consider feature reduction. However, it can easily be computationally expensive to many features. The principal component analysis (PCA) proposed by Jolliffe [24] utilizes a classical statistical method for extracting a lower dimensional space. This is carried out by examining the covariance design of multi-variate statistical inspections. Yet, the main idea behind PCA is keen to ensure that the new features are of high quality. This provides the ability to describe as much of the overall variation in the data as possible. Such a target sits well with the eigenvalue decomposition to perform on the covariance matrix. The analytical steps are to ensure that the measure of covariance between features is consolidated:

$$f_{PCA} \approx \text{eig\_decomp}\left(V = \sum_f (f - m)(f - m)^T \right). \qquad (5.8)$$

Here, f is the features extracted by above-mentioned CNN, and m denotes the mean vector of the features. By considering variation around the mean vector on each pair of the features, the covariance V matrix is obtained. Then the eigenvalues and eigenvectors of V are calculated to sort the eigenvalues in descending order. As a consequence, the actual transition matrix can build by choosing the preferred number of components from the eigenvectors. By multiplying the initial feature f_i with the derived transition matrix, it allows to get a lower dimensional representation. Since the required cumulative percentage of variance is interpreted by the principal axes, the percentage can be used as a threshold to define the number of the most principal components to choose. So, the PCA method uses the covariance calculation and therefore is computationally rather expensive [25].

We overcome this by exploiting the analysis of features in relation with its response to class to seek an effective feature reduction method. Thus, moving the learning attention to the responding capacity of each feature f to a particular class c would require consideration of the joint distribution $p(f,c)$ in the training set of samples:

$$p(f,c) = p(f|c) p(c). \qquad (5.9)$$

The key point is the use of image sample analysis allowing for common signatures of feature behavior and their individual behavioral patterns to be established. By denoting s for an image sample, $f(a)$ is the value of the feature f extracted for the image sample. Therefore, in order to identify from eqn (5.9), it is important to consider the conditional probabilities $p(f \mid f(a))$ and $p(f(a)|c)$ for each image sample

$$p\big(f \mid f(a), c\big) \approx p\big(f|f(a)\big) p\big(f(a)|c\big). \tag{5.10}$$

This helps the expanding of $p(f,c)$ to the vision level of samples to rate the joint probability:

$$p(f,c) = \sum_a p\big(f|f(a)\big) p(f(a) \mid c) p(c). \tag{5.11}$$

It is clear that

$$p(f \mid c) = p(f,c) / p(c). \tag{5.12}$$

As a consequence, the condition probability of a feature on class can be determined through checking the training set of image samples:

$$p(f \mid c) = \sum_a p\big(f|f(a)\big) p(f(a) \mid c). \tag{5.13}$$

Particularly, given an image sample $\hat{a}$ and its feature value $f(\hat{a})$, a simple way to estimate $p(f(\hat{a})|f(a))$ for eqn (5.13) is the ability of applying discrete transformation by a function d for the feature and then using comparison of the values [26]:

$$p\big(f\big(\hat{a}\big) \mid f(a)\big) = \big(d\big(f\big(\hat{a}\big)\big) == d\big(f(a)\big)\big). \tag{5.14}$$

Note that the frequency of appearance of feature value $f(a)$ for a particular feature f and a specific sample a regard to a set of samples having an assigned class c can offer evaluation of $p(f(a)|c)$, which is an important member of formula (5.13). So the estimation $p(f|c)$ of is now enabled.

In the presence of available estimation of conditional probability $p(f|c)$ of a feature, given a class (5.13), the prior probability $p(f)$ can be achieved by checking all the classes as follows:

$$p(f) = \sum_c p(f \mid c) p(c). \tag{5.15}$$

Since the frequency of appearance of particular in the training dataset can be resumed, estimation of $p(c)$ is feasible [see eqn (5.15)]. After having prior distribution for all features, we are left with sorting the result to select the features remarked f by whose values are above a pre-determined threshold.

5.3.3 Classification

By its nature, feature extraction and feature reduction are significant tasks for our classification. Figure 5.2 illustrates the role of feature and its reduction version in the reference schema: image feature f is determined by CNN on the original chest X-ray image and then feature is reduced to have $\dot{f}$.

Note that to get the condition probability of a reduced feature on a class, checking training set of image samples allows making (5.4) achievable by the following equation:

$$p\left(\dot{f} \mid c\right) = \sum_{a} p\left(\dot{f} \mid f(a)\right) p(\dot{f}(a) \mid c).$$ (5.16)

In particular, since we previously allowed access to the training data, the prior $p(c)$ for each class c can be estimated by the data with appearance of a pair of reduced features $\dot{f}(a)$ and assigned class for each image sample a:

$$p(c) = \sum_{a} p\left(c \mid f(a)\right) p\left(\dot{f}(a)\right).$$ (5.17)

So far, we have looked at feature extraction and reduction for chest X-ray images. Having defined $p(\dot{f} \mid c)$ by eqn (5.16) and $p(a)$ by eqn (5.17) for a test chest X-ray image a, a class can be estimated by a practicable form of eqn (5.4). The most likely class c now is associated with the features $\dot{f}$ of the image a:

$$c_{\text{MAP}} = \text{argmax}_{c \in C}\, p(\dot{f}(a) \mid c) p(c).$$ (5.18)

In practice, of course, we conduct the formula (5.18) by implementing the support vector machines (SVMs) [27] that search the maximum-margin hyper-plane to separate samples of different classes.

By comparison, the estimated class and ground truth class, true positive (TP), true negative (TN), false negative (FN), and false positive (FP) are summed up and make accuracy evaluation [28] possible:

$$\text{Acc} = \frac{\text{TP} + \text{TN}}{\text{TP} + \text{TN} + \text{FP} + \text{FN}}.$$ (5.19)

5.3.4 Algorithm

The described inference learning method is implemented in programming schema with main modules consisting of feature extraction, feature reduction, training, and testing. Figure 5.2 displays the training tasks by blocks

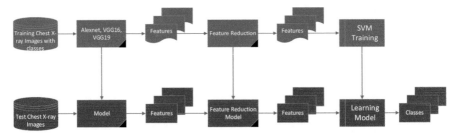

Figure 5.2 Proposed schema of classification for chest X-ray images.

with blue background color. The feature extraction module is now performed by one of the CNNs including AlexNet, VGG16, and VGG19. The feature extraction model is used later for test images, shown by a block with dark yellow background color. Typically, feature reduction model produces selected features for training and test data. Finally, the SVM are applied for classification.

In Figure 5.3, we denote major steps for the algorithms of the describer method. Given a set of chest X-ray image for training, feature extraction is performed with explanation of formulas (5.5) and (5.7). Then, the feature extraction is followed with mentioned formulas (5.14) and (5.15). The training task by SVM is clarified by formula (5.17).

We use delivered models from training task (and _SVM) to carry out main steps in the test.

5.4 Experimental Results

The chest x-ray images we used has 100 frontal images that are parts of the COVID-19 Radiography Database [29]. In fact, there are three classes pre-determined for the dataset containing normal, Covid-19, and viral pneumonia. The number of images is shown in Table 5.1, and the distribution of classes on the data set is illustrated by Figure 5.4.

In order to detect the class of Covid-19, we group the classes of normal and viral pneumonia. In order to detect the class of Covid-19, we group the classes of normal and viral pneumonia into a negative class for a binary classification. Additionally, we use a cross-validation method with five folds and each fold has a split of 80% images for training and 20% for test.

In regard to the image size, we resize the images to 224 × 224 for VGG16 and VGG19, whilst use the size of 227 × 227 for AlexNet. For each CNN model, features of the output layer in the form of an array with 1000 real numbers are extracted from the training set and test set. Then a min−max

Input training set of chest X-ray images $\{a, c\}$

Input test set of chest X-ray images $\{\hat{a}\}$

Output class $\{\hat{c}\}$

Begin

// Training

$[Rf, model\_FE]$ = Feature_Extraction_by_CNN (a)//(5,7)

$[Rf, \dot{model}\_FR]$ = Feature_Reduction (f) //(14-15)

$model\_SVM$ = SVM_Learning $(\dot{f}, c)$ //(17)

// Test

f = Feature_Extraction_by_CNN $(\hat{a}, model\_FE)$ //(5,7)

$\dot{f}$ = Feature_Reduction $(f, model\_FR)$ //(14-15)

$\hat{c}$ = SVN_Classify $(\dot{f}, model\_SVM)$ //(16,18)

End

Figure 5.3   Algorithm of classification for chest X-ray images.

Table 5.1   Dataset for the experiments.

Class	Image database [29]	Images from [29] for the case study
Normal	10,192	338
Covid-19	3516	117
Viral pneumonia	1345	45
Total	15,053	500

scaling is applied to the training set for an interval of [R.05,95] to reserve space for values of features in test dataset, which could be out of the min–max values of the training set. For each min–max scaled feature set, the feature responding to classification was performed.

Basing on that the features were selected by 10%, 20%, etc. to 100% of total 1000 features for SVM binary classification. As accuracy can be estimated for each test sample, averaging of each cross-validation with five splits allow calculating averaged accuracy. In Table 5.2, we printed the accuracy for

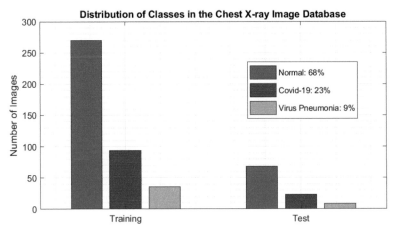

Figure 5.4 Distribution of classes in the chest X-ray image database.

Table 5.2 Performance report by accuracy.

Method/ reduction	10%	20%	30%	40%	50%	60%	70%	80%	90%	100%
VGG16	0.841	0.877	0.871	0.877	0.890	**0.917**	0.909	0.892	0.884	*0.867*
VGG19	0.851	0.877	0.909	0.905	0.911	**0.917**	0.909	0.905	0.896	*0.908*
AlexNet	0.811	0.831	0.892	0.886	0.909	**0.897**	0.888	0.913	0.883	*0.896*

classification without feature reduction by italic numbers: VGG16 gets 0.867, VGG19 has 0.908, and AlexNet takes 0.896. So, VGG19 is the best in this case.

In the following, we look at a reduction rate of 50% where scores are underlined. It is clear that the scores are increased with 0.890 for VGG16, 0.911 for VGG19, and 0.909 for AlexNet. However, the best scores are shown up in the column of reduction ratio of 60% with print in bold.

The achieved accuracy for VGG16 and VGG19 is 0.917, and 0.897 is the score of AlexNet. As such, the feature reduction proved its effect on performance enhancement for our case study with a solution combining CNN, SVM, and feature engineering. The graphs presented in Figure 5.5 are designed to show which reduction rate should be chosen for each CNN model. In other words, each bar shows the accuracy scores and high bars are preferred to select.

The feature reduction decreases the use of computing resources covering time and disk space. However, the reduction usually decreases also the performance. To be able to keep and enhance the prediction accuracy, features need a careful analysis and experiments for consolidating the validation of the method.

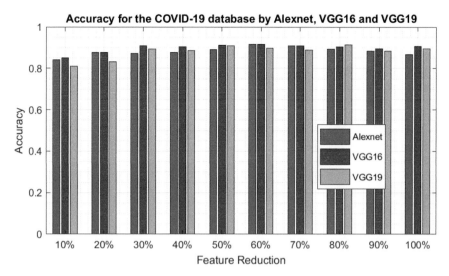

Figure 5.5 Distribution of classes in the chest X-ray image database.

In the following, we present related works in Table 5.3, where databases used for experiments and reported accuracy are shown in columns. The DenseNet169 [4] is addressed to images from database of COVID-CT [32] that was provided by CT images with assistance of masks, allowing an accuracy of 89%. A solution by a combination of ResNet50 and MobileNet [5] was tested on images selected from two databases [15] and [30] giving accuracy of 84.35%.

An inception migration neural network [8] with images from a pneumonia CT dataset achieved 82.90%. So far, the variation of methods and databases can be seen in the table as research efforts and interest spent for the Covid-19 chest X-ray image recently. Overall, this means that our method is good at the classification and this contributes one more solution for the particular question of diagnosis with medical image analysis.

5.5 Discussion and Conclusions

We have described an integrated solution of CNN and SVM for detecting Covid-19 symptoms or defection from lung area of chest X-ray images. The method is capable of dealing with medical images that typically contain complex objects. We found that CNN models including VGG16, VGG19, and AlexNet can offer significant features for classification. One of the most important factors that this chapter emphasizes is the need to analyze CNN-based features to improve computational performance in terms of both timing

Table 5.3   Performance report of related work.

Method	Image database	Acc (%)
DenseNet169 [4]	COVID-CT [31] with masks	89.00
Combination of ResNet50 and MobileNet [5]	Images from [15] and [30]	84.35
Inception migration neuron network [8]	Pneumonia CT [31]	82.90
ResNet-101 [10]	Two databases from [11]	77.30
Five convolutional layers CNN [12]	200 images from [15] and [29]	90.64
Influence of batch size and image size [16]	Image from COVID-19 Radiography Database [29]	87.00
Dual sampling attention network ResNet34 [18]	4982 images [18]	87.90
ResNet and location attention [19]	175 healthy, 224 viral pneumonia, and 219 Covid-19 [19]	86.70
Integrated VGG16&SVM&50% (our)	500 images from [29]	89.00
Integrated VGG19&SVM&50% (our)	500 images from [29]	91.11
Integrated AlexNet&SVM&50% (our)	500 images from [29]	91.09

and accuracy. For the three mentioned CNN models, a feature reduction rate of 60% may be preferable. In other cases, a combination of theoretical and experimental approaches for feature extraction and feature reduction will yield superior performance. At the moment, our data was selected for a limited number of images due to the limitation of our computer capacity. In the future, we plan to enlarge the number of images required for the study to better learn latent features of medical images. This will allow us to realize our method in a practicable environment. Another constraint of our work that we want to deal with is to examine a larger number of potential CNN models for extracting latent features.

References

[1] Ahmed Elnakib, Hanan M. Amer, Fatma E.Z. Abou-Chadi, (2020) Early Lung Cancer Detection using Deep Learning Optimization, International Journal of Online and Biomedical Engineering (iJOE) Vol 16, No 06, pp. 82–94

[2] Hamed S. G., Yaghoub Pourasad, (2020) Clustering of Brain Tumors in Brain MRI Images based on Extraction of Textural and Statistical

Features, International Journal of Online and Biomedical Engineering (iJOE) Vol 16, No 14, pp. 95–106

[3] Nanditha B R, Geetha Kiran A, Chandrashekar H S, et al., (2020) Oral Malignancy Detection Using Color Features from Digital True Color Images, International Journal of Online and Biomedical Engineering (iJOE) Vol 16, No 14, pp. 95–106

[4] J. Zhao, Y. Zhang, X. He X., P. Xie, (2020). COVID-CT-Dataset, A CT Scan Dataset about COVID-1, arXiv preprintarXiv, 2003.13865

[5] Ritika Nandi, Manjunath Mulimani (2021) Detection of COVID-19 from X-rays using Hybrid Deep Learning Models, DOI, 10.21203/rs.3.rs-468236/v1, Research Square

[6] Kaiming He, Xiangyu Zhang, Shaoqing Ren, and Jian Sun. (2016) Deep residual learning for image recognition. 2016 IEEE Conference on Computer Vision and Pattern Recognition (CVPR), Jun 2016. doi, 10.1109/cvpr.2016. 90. URL http://dx.doi.org/10.1109/cvpr.2016.90

[7] Andrew G. Howard, Menglong Zhu, Bo Chen, Dmitry Kalenichenko, Weijun Wang, Tobias Weyand, Marco Andreetto, and Hartwig Adam. (2017) Mobilenets, Efficient convolutional neural networks for mobile vision applications, 2017.

[8] Shuai Wang, Bo Kang, Jinlu Ma, Xianjun Zeng, Mingming Xiao, Jia Guo, Mengjiao Cai, Jingyi Yang, Yaodong Li, Xiangfei Meng, Bo Xu, (2020) A deep learning algorithm using CT images to screen for Corona Virus Disease (COVID-19) European Radiology doi, 10.1007/s00330-021-07715-1

[9] A. Majkowska, S. Mittal, D. F. Steiner et al., (2020) Chest radiograph interpretation with deep learning models, assessment with radiologist-adjudicated reference standards and population adjusted evaluation, Radiology, vol. 294, no. 2, pp. 421–431.

[10] Mohd Zulfaezal Che Azemin, Radhiana Hassan, Mohd Izzuddin Mohd Tamrin, and Mohd Adli Md Ali 4, (2020), COVID-19 Deep Learning Prediction Model Using Publicly Available Radiologist-Adjudicated Chest X-ray Images as Training Data, Preliminary Findings, Hindawi International Journal of Biomedical Imaging, Volume 2020, Article ID 8828855, 7 pages, https://doi.org/10.1155/2020/8828855

[11] Kaiming He, Xiangyu Zhang, Shaoqing Ren, Jian Sun, (2015), Deep Residual Learning for Image Recognition, arXiv.org > cs > arXiv, 1512.03385

[12] Faizan Ahmed, Syed Ahmad Chan Bukhari, and Fazel Keshtkar (2020) A Deep Learning Approach for COVID-19 & Viral Pneumonia Screening with X-ray Images. Digit. Gov., Res. Pract. 2, 2, Article 18 (December 2020), 12 pages., https://doi.org/10.1145/3431804

[13] Muhammad E. H. Chowdhury, Tawsifur Rahman, Amith Khandakar, Rashid Mazhar, Muhammad Abdul Kadir, Zaid Bin Mahbub, Khandaker Reajul Islam, Muhammad Salman Khan, Atif Iqbal, Nasser Al-Emadi, Mamun Bin Ibne Reaz, and T. I. Islam. (2020). Can AI help in screening Viral and COVID-19 pneumonia? Retrieved from https://www.kaggle. com/tawsifurrahman/covid19-radiography-database.

[14] Nabeel Sajid. (2020). COVID-19 Patients Lungs X Ray Images 10000. Retrieved from https://www.kaggle.com/nabeelsajid917/ covid-19-x-ray-10000-images.

[15] Joseph Paul Cohen, Paul Morrison, and Lan Dao. 2020. COVID-19 image data collection. Retrieved from https//github.com/ieee8023/ covid-chestxray-dataset.

[16] Ashura Binti Hasmadi, Mehak Maqbool Memon, Manzoor Ahmed Hashmani, (2020) Interactive Automation Of COVID-19 Classification Through X-ray Images Using Machine Learning, Journal of Independent Studies and Research - Computing, 10.31645/10

[17] Tho D.X., Anh D.N. (2021) Imbalance in Learning Chest X-ray Images for COVID-19 Detection. In, Soft Computing, Biomedical and Related Applications. Studies in Computational Intelligence, vol 981. Springer, Cham. https://doi.org/10.1007/978-3-030-76620-7_9

[18] Xi Ouyang, Jiayu Huo, Liming Xia, Fei Shan, Jun Liu, Zhanhao Mo, Fuhua Yan, Zhongxiang Ding, Qi Yang, Bin Song, Feng Shi, Huan Yuan, Ying Wei, Xiaohuan Cao, Yaozong Gao, Dijia Wu, Qian Wang, Dinggang Shen, (2020) Dual-Sampling Attention Network for Diagnosis of COVID-19 From Community Acquired Pneumonia. IEEE Trans. Medical Imaging 39(8), 2595–2605 (2020)

[19] Xu X., Jiang X., Ma C., Du P., Li X., Lv S. (2020). Deep Learning System to Screen Coronavirus Disease 2019 Pneumonia. arXiv preprint arXiv, 200209334. [RGoogle Scholar]

[20] Barber, D. (2012) Bayesian Reasoning and Machine Learning, Cambridge University Press.

[21] Krizhevsky, Alex, Sutskever, Ilya, Hinton, Geoffrey E. (2017). ImageNet classification with deep convolutional neural networks (PDF). Communications of the ACM. 60 (6), 84–90. doi, 10.1145/3065386. ISSN 0001-0782. S2CID 195908774

[22] Karen Simonyan, Andrew Zisserman, (2015), Very Deep Convolutional Networks for Large-Scale Image Recognition, arXiv, 1409.1556 [Rcs.CV]

[23] Dario Garcia-Gasulla Dario, Garcia-Gasulla Ferran, Parés Ferran, Parés Armand, Vilalta Armand, Vilalta Show, Toyotaro Suzumura, (2017), On the Behavior of Convolutional Nets for Feature Extraction March

2017Journal of Artificial Intelligence Research 61 Follow journal DOI, 10.1613/jair.5756

[24] Jolliffe, I.T. (1986) Principal Component Analysis. Springer, New York, NY, (1986)

[25] Oza, N.C., Tumer, K. (1999) Dimensionality Reduction Through Classifier Ensembles. Technical Report NASA-ARC-IC-1999-124, Computational Sciences Division, NASA Ames Research Center, Moffett Field, CA, (1999)

[26] Dao Nam Anh, (2021) Manifold Based Data Refinement for Biological Analysis. In, Kreinovich V., Hoang Phuong N. (eds) Soft Computing for Biomedical Applications and Related Topics. Studies in Computational Intelligence, vol 899, pp 141–152. Springer, Cham. https://doi.org/10.1007/978-3-030-49536-7_13. Print ISBN 978-3-030-49535-0, Online ISBN 978-3-030-49536-7

[27] Ben-Hur, Asa, Horn, David, Siegelmann, Hava, Vapnik, Vladimir N. (2001), Support vector clustering, Journal of Machine Learning Research. 2, 125–137.

[28] Powers, David M. W. (2015). "What the F-measure doesn't measure". arXiv, 1503.06410

[29] COVID-19 Radiography Database https://www.kaggle.com/tawsifurrahman/covid19-radiography-database

[30] Daniel S. Kermany, Kang Zhang, and Michael H. Goldbaum. (2018) Labeled optical coherence tomography (oct) and chest x-ray images for classification.

[31] J. Zhao, Y. Zhang, X. He X., P. Xie, (2020). Covid-CT-dataset, a CTt scan dataset about Covid-19, arXiv preprintarXiv, 2003.13865

6

Predicting Genetic Mutations Among Cancer Patients by Incorporating LSTM with Word Embedding Techniques

P. Shanmuga Sundari[1], J. Jabanjalin Hilda[2], S. Arunsaco[3]* and J. Karthikeyan[4]

[1]Department of Computer Science and Engineering (Data Science), SVCET (A), India.
[2]Jabanjalin Hilda, School of Computer Science and Engineering, VIT University, India.
[3]Department Mechanical Engineering, SVCET (A), India.
[4]Department of Humanities and Science, SVCET(A), India.
Email: sigashanmu@svcet.in.; jabanjalin. hilda@vit.ac.in; jk4english@gmail.com
*Corresponding author's email: arunsaco@svcet.in

Abstract

In recent years, technological developments support advanced treatments for dangerous diseases like cancer and assist in life saving. The cancer tumor has thousands of genetic mutations. Personalized medicine involves the systematic study of genetic mutation and other related information. Understanding the cancer tumor growth is a challenging task even when advanced genetic analysis is adopted. At present, understanding genetic mutation is done manually. Advance techniques like machine learning provide the way to find the genetic mutation growth automatically. To automate the process, some studies used classification algorithms like random forest, naive Bayes, XGBoost, and LSTM. The above-mentioned classification algorithms are plagued by issues of less accuracy due to spare data. The computation cost is high to train the model for prediction. To automate the process, the proposed method integrates the LSTM model with word embedding word2vec technique. The LSTM-based neural network model predicts the gene sequence and increases the classification accuracy by using traditional recurrent neural networks. The "long short-term memory (LSTM)" model received recent popularity among neural network models. Word embedding techniques convert the text into machine understandable code like vectors. These vectors will easily adapt to machine learning models. The proposed method combines the genetic variations along with clinical text. Word embedding techniques are mainly used to understand the semantic meaning of clinical annotations, which will enhance performance. The experimental result reveals that the LSTM-based word embedding technique achieves 84% accuracy. The result analysis of the proposed combined approach outperforms the existing methods.

6.1 Introduction

Advanced technology provides the way to give personalized treatment for patients and provides personalized medicines (without side effects and with in-depth analysis of the gene sequence) to increase the recovery rate, fasten the treatment, and to safeguard life. In 2008, the World Health Organization (WHO) recognized 7.6 million deaths worldwide due to cancer (13% of total global mortality), with 70% of the deaths occurring in average income countries. In 2012, more than 23 million people are affected by malignant tumors according to the WHO, while projections for 2030 envisage more than 13 million cancer-related deaths for the entire world population. According to the WHO report, there are three main factors involved in the increase of cancer, namely: heredity, wider environment, and behavior.

The public health challenge related to cancer therefore seems to us all the more important that, at the global level, the same three factors, sometimes combined, explain the origin of cancers: heredity (responsible for 5% of cases), the wider environment (pollution, occupational exposure, etc.), and behavior (smoking, alcohol, obesity, etc.). Cancer can be considered as the disease of the early 21st century, as cardio-vascular diseases are of the 20th century and infectious diseases of the 19th century. Cancer deaths are more common in people over 65, diagnosed with 58% new cases each year and accounting for three-quarters of deaths worldwide. The survival rate also decreases with age: observed over five years, this rate is 52% on average, while it drops to 39% among people aged 75 and over. The aging of the population leads to an increase in the number of cancers.

Advances in molecular biology and genome knowledge are revolutionizing the cancer treatments. The symptoms and causes may vary between individuals. No two cancer symptoms are identical. The nature of the genetic compositions is mutated for each cancer type. Some of these genetic differences are the "drivers" of tumor growth. The principle of personalized medicine is to treat by targeting these specific genetic differences that affect the development and evolution of cancer. Advanced technology used in cancer studies provides the way to establish a person's unique combination of genes, which will be used to find the corresponding drugs for the cancer.

The method is based on scientific advances to understand how a person's unique genetic and molecular sequences make them subject to certain diseases. If cancer is considered a major public health issue, the implementation of these innovative and expensive therapies cannot be done without thinking about the cost of the treatments, particularly for health insurance.

Recent advancement in technology helps to provide better treatment for cancer patients. Interdisciplinary research provides the way to understand the insights of the causes of diseases, which will help the doctors to provide medicines based on the natural phenomena. Technologies like IoT, machine learning, and deep learning methods are used to analyze patients' history, which will give hidden knowledge from the data [26–28].

This chapter proposes a novel method by adopting deep learning technique with word embedding method that analyzes the patients' electronic health record along with semantic analysis on clinical text data, which will produce a better understanding of the causes of the disease. Deeper understanding and knowledge helps in giving better treatment for the patients and providing personalized treatments. The proposed method was tested with earlier methods and the results are observed. The observed results show that the proposed method outperforms compared to existing methods.

6.2 Related Work

6.2.1 Basic feature engineering

The bag-of-words represents the documents by the frequencies of each word it contains. However, if a term appears throughout the corpus very often, it means that it does not carry meaningful information about a particular document. The TF-IDF (term frequency-inverse document frequency) model followed the idea that a word that frequently occurs in one document but not so frequently in the corpus may be important to represent this document. Therefore, it considers the "inverse document frequency (IDF)" along with "term frequency (TF)." Similar to the bag-of-words model, the drawback is that it focuses on every word's occurrences but does not consider semantic similarities and orders. To overcome the above-mentioned drawback, the author [1, 2] presented a new word embedding method, word-2vec, that learns the number representation for every word using a partial neural network language method. In this method, word vector is trained to maximize the log probability of its adjacent words. Therefore, the embedded word vectors hold the context information and enable us to measure the similarities among words. doc2vec [3, 25] is a framework used to find the document similarity measure. It is quite similar to word2vec but trained a paragraph vector and the word vectors to predict the words in the documents. This paragraph vector further can be used in classification. Another method that is focused on circulating the distance between two documents is called word mover distance [4, 24]. It is based on the word vectors and follows the idea of training a model to get the minimum cost for "travel" from one document. Support vector machines and latent Dirichlet allocation (LDA) have been used for certain tasks, such as the classification of patient notes [5] or other documents in diseases such as diabetes that have shown satisfactory results [6, 7, 23].

Literature studies from many various NLP word embedding stated that this feature extraction techniques have received extensive impact on feature extraction but, further testing is still required before they can be effectively applied in the biomedical field. They adjust different hyper-parameters including the sub-sampling, minimum-count, context window size, etc. And they find that sentence-shuffled text corpora with proper size and parameter setting show better results. And skip-grams generally show better result than CBOW.

The study [11] on Cancer Hallmark Text Classification using "Convolutional Neural Networks (CNN)" they have proposed using a convolutional neural network (CNN) method to tackle a similar problem of feature extraction method when compare to support vector machine. The original problem in this chapter is to classify text of abstracts of biomedical publications according to ten cancer hallmarks. Hallmarks of cancer are different biological properties that can lead to cancer. Each abstract can be classified to more than one hallmark, or not classified. So, the original problem is a multi-class problem, whereas the problem of this study contains a multi-class problem, i.e., each text input is classified as one result type. Baker *et al.* tested their CNN model and obtained competitive results compared to SVM models. The tuned CNN model has outdone the two SVM models presented in the study. The advantages of this approach are that it needs fewer manual engineering. General SVMs need a good amount of knowledge and engineering such as analyzing useful features (mentioned above) from the texts to obtain good learning results. Some of the drawbacks of using CNN in this type of NLP problem include a large amount of computation and large number of optimization choices for parameters.

6.2.2 Classification method

Neural networks require a large amount of training data and they can result in overfitting. The WDM works well in clustering, but it also suffers from its complexity and sometimes it does not work well with classification problems. Several other attempts have been made to fix this. Besides, due to its feedback structure, RNN can naturally use the sequence information widely used in NLP. LSTM fixes this problem by introducing the "forget" gate to what extent a value remains in memory. Consider SVM and XGBoost, which are employed better for small-scale dataset without overfitting. Suppose there is training data, each one was classified into two categories. SVM algorithm constructs a classifier that assigns new values to one or the other category, making it a binary classifier that is not probabilistic. Since for SVM, only support vectors but not all the samples are used to construct the separating hyper-plane that is the classifier. However, to be more scientific, we need to further perform generalization tests (also for SVM with Gaussian and linear kernel).

6.3 Data Overview

6.3.1 Dataset structure

The dataset is taken from MSKCC, which contains genetic variations and text. There are separate files for each dataset. The features are data ID, the gene names, the variation type, and the cancer type (class label). In another file, the data ID is matched with the file, which will provide the clinical annotation in plain text. The files are combined with features with a class label.

6.3.2 Data pre-processing

The datasets contain some categorical features. The feature engineering steps are followed to convert the categorical data into numerical form.

6.3.3 Most frequent genes and class

Some data analysis has been done with the help of visualization techniques to find the most frequent genes and class. Figure 6.1 represents the most frequent genes and their class labels. The observed results show that gene frequency distribution varies a lot, such as for class 1, TP53 and BRCA1 are much more frequent than other genes. Similarly, top few genes take up a big percentage in class 5, class 6, and class 9, indicated by the great frequency gap between the

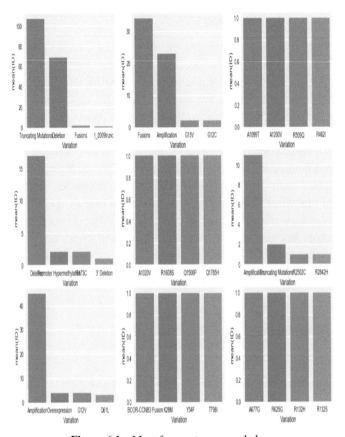

Figure 6.1   Most frequent genes and class.

top 1 gene and the top 10 gene. While the gene distribution seems more balanced in class 2, class 3, and class 7. Moreover, the most representative group of genes is distinctively different for each class. The most appearing genes for nine classes are TP53, KIT, BRCA1, PTEN, BRCA1, BRCA2, EGFR, BCOR, and SF3B1, respectively, with the only overlap being class 3 and class 6. But deeper understanding reveals that the two classes have a very different combination of first-tier genes but also the distribution of them. The overall study reveals that the gene looks like a very promising predictive factor for the label.

6.3.4 Most frequent variation and class

Unusually, variations are much more diverse and the frequency of the top variation is much lower than the top gene. This is because only a limited number of genes (around 20,000 for human beings, among which only less than 10% are cancer-related), but almost unlimited possibility of the number of variations. (As we have discussed before, a variation is basically a form of gene which is different from wild-type genes. For a specific gene, even assuming it is composed of only 500 amino acids, the possible point mutations only could be as many as 500*20, not to say deletions, insertions, and many other forms of variation.)

Some classes like class 3, class 4, and class 9 do not even have a single repeated variation, indicated by the top frequent one having only one count. Top variations in other classes, such as truncating mutations in class 1, deletion in class 4, and amplification in class 7, are aggregated forms of mutations, meaning themselves do not refer to any specific variation, unlike A1099T, a point mutation which specifically refers to a point mutation from Ala to Thr in 1099 position of this gene, truncating mutations and deletions is a joint name in their cases. Therefore, even variation also shows some correlation toward class.

6.3.5 Text length distribution and class

Text length is created by encoding the character length of the text field. Remarkably, the distributions of text length are very heterogeneous. Class 3 (label 2.0), for example, has a very narrow distribution, in which almost 95% of its text is in range (20 and 80 K). In contrast, class 8 and class 9 have expanded text length distribution, indicated by the long and narrow spindle shape, which is represented in Figure 6.2.

6.3.6 Word cloud to visualize data

To make data easier for the human brain to grasp and draw conclusions from, data visualisation is the practise of putting information into a visual context, like a map or graph. Data visualization's major objective is to make it simpler to spot patterns, trends, and outliers in big data sets. The most frequent words in text are mutat, cell, cancer, tumor, active, and gene. Figure 6.3 shows the word cloud representation of the frequent terms that are present in the clinical annotation.

6.4 Methodology

The proposed method contains three major components: pre-processing and feature extraction, word embedding, and classifier, as shown in Figure 6.4.

6.4.1 Pre-processing and feature extraction

Text pre-processing is the first step in a text classification task. It involves converting the upper case to lowercase, removing numbers, removing

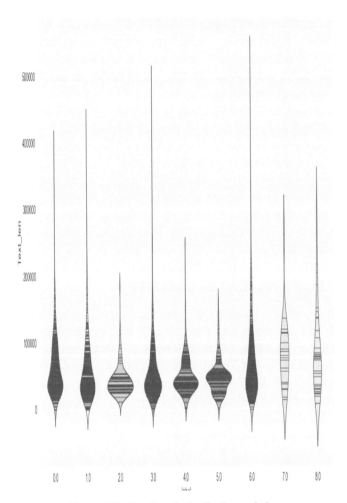

Figure 6.2   Text length distribution and class.

punctuations, and removing stop words. After that, we will perform tokenization, and then we use lemmatization to chop off the affixes for better accuracy.

6.4.2 Word embedding

Word embedding method is used to find the vector representation of the text from the vocabulary list of corpus data. The vector representation can be either text classification using a neural network or clustering based on text statistics. There are three main layers adapted to learn the word embedding from the text. They are embedding layer, word2vec, and doc2vec.

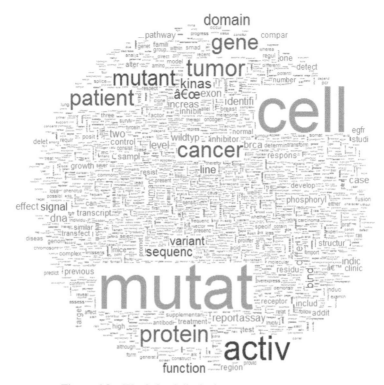

Figure 6.3 Word cloud displaying most common terms.

6.4.2.1 Embedding layer

In this layer, the word embedding that is the language modeling was learnt with the help of neural network model. The pre-processed that is the cleaned data is required for "one-hot" encoding representation. The "one-hot" encoding is a matrix representation of the vocabulary list. The maximum word length is used to fix the size of the vector space. Initially, random numbers are used to represent the vectors. The embedding layer uses the neural network model to fit the model based on the supervised learning method using the back-propagation algorithm.

6.4.2.2 Word2Vec

Two models were used to learn the word embedding, which are "continuous bag-of-words, or CBOW model, and continuous skip-gram model" [3]. The continuous bag-of-words model predicts the current model based on the text contexts associated with the desired word. The "skip gram model" predicts the neighboring words based on the current word [3]. In both the models, the

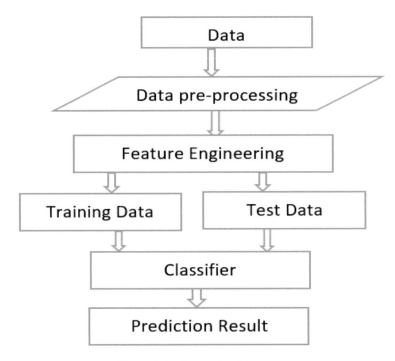

Figure 6.4   Proposed architecture.

learning can be done with the help of text contexts. The situation is learnt by window of adjacent terms. The frame ("window") is a learning parameter.

6.4.2.3 GloVe
In this layer, the document can be converted in the form of vectors. The GloVe is another word embedding method. It uses several techniques and equations to make an embedding layer [16]. Let us consider the following equation:

$$\left. \begin{array}{c} Y_{ij} - matrix\ Y\ represent\ the\ number \\ of\ times\ word\ j\ occurs\ in\ the\ context\ of\ word\ i \\ Y_i = \Sigma_k\ Y_{ik} \\ P_{ij} = P(j\,/\,i) = Y_{ij}\,/\,Y_i \end{array} \right\}. \qquad (6.1)$$

The ratio of co-occurrence of probabilities is given in the following equation:

$$F(t_i, t_j, t_k)$$

$$= \frac{P_{ik}}{P_{jk}} \qquad (6.2)$$

where t_i is term $\in R^d$ are term vectors, t_k – probe term, t_i and t_j is the term correlation between t_i and t_j, and $\dfrac{P_{ik}}{P_{jk}}$ cooccurance probabilities of term t_i and t_j.

It maintains the linear relationship among the vectors. In the term–term co-occurrence matrix, the non-zero elements are modeled with the help of eqn (6.2) using statistical inferences.

6.4.3 Classifier

As a part of classification methods in this study, different classifiers are used, namely naïve Bayes, random forest, XGBoost, voting classifier, and LSTM.

6.4.3.1 Naive bayes

Naïve Bayes models work under the principles of conditional probability. The model predicts the probability of an instance belonging to a class with a given set of feature value. It is a probabilistic classifier. It is called naïve because it assumes that one feature in the model is independent of the existence of another feature. In other words, each feature contributes to the predictions with no relation between each other [22].

6.4.3.2 Random forest

Random forest classifier is the most popular classification algorithm in machine learning techniques. Random forest classifier is the combination of decision trees that are completely independent to each other. The mode predicts the new classification based on the voting results of each classifier and class label of the model. The following steps illustrate the key idea to budding the random forest classifier.

- It identifies the exact value of variable D, which denotes the number of features of every feature subset.

- Select a new feature subset n_k from the random-based feature of the whole feature set D. The new set n_k is a sequence of independent feature sets denoting $n_1, n_2, \ldots, n_k$.

- The training dataset is considered a training set group with a feature subset to construct the decision tree. The function $n(Y, n_k)$ (where Y denotes the sample inputs) represents every classifier.

- The new n_k subset is selected by repeating the above-mentioned steps until all the feature subsets must be traversed. The random forest classifier is implemented.

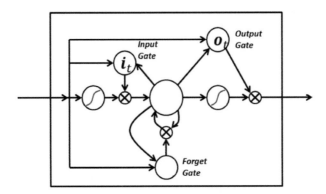

Figure 6.5   LSTM architecture.

- Enter the test model, according to the voting results of each category, and the label of a certain category in the sample is determined.

6.4.3.3 XGBoost

XGBoost algorithm is an ensemble technique in which weak models are added to correct the errors made by them. New models will predict the errors of prior models and then added together to create new models. XGBoost algorithm is faster than other algorithms. It automatically takes care of missing values. It incorporates a new sparsity-aware algorithm to handle different types of sparsity patterns in the data. It uses multiple cores on the CPU to enable parallel computing.

6.4.3.4 LSTM

"Long short-term memory (LSTM)" networks are based on memory cell. Memory cell can store the information for long time or short time depending on the importance of information. Each LSTM memory cell contains three gate types: "input gate, forget gate, and output gate." The "input gate" allows the information to enter the cell. The "forget gate" permits the information to be forgotten or vanished from the cell. The newly generated information will occupy the vanished space. The "output gate" allows the output information from the cell, which is illustrated in Figure 6.5.

6.5 Experiments

This section details how the proposed approach is applied to the dataset. Dataset is split up to train and validation data. Baseline data is experimented with several classifiers like logistic regression, random forest, naïve Bayes,

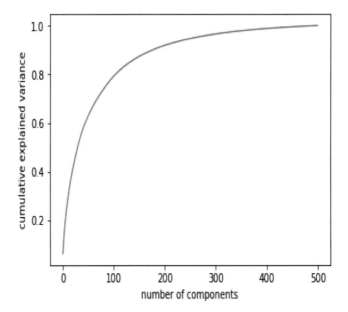

Figure 6.6 Principle component analysis graph.

Table 6.1 Evaluation metrics results.

Models	Accuracy	Precision	Recall	*F*1-score
Naïve Bayes	0.63	0.59	0.58	0.57
Random forest	0.65	0.58	0.56	0.56
XGBoost	0.62	0.56	0.56	0.56
Voting classifier	0.67	0.60	0.58	0.59
LSTM	0.84	0.80	0.82	0.81

XGBoost, and voting classifiers. Word2Vec is used for all the classifiers. The LSTM networks are trained (fine-tuned) using labeled data with back-propagation. LSTM is used with word embedding called word2vec. Several input features are identified with the help of Figure 6.6.

6.6 Results

On test data LSTM gave the highest micro-averaged precision, recall, and *F*1-score. The results are tabulated in Table 6.1. LSTM has outperformed all the algorithms with good accuracy. The results are tabulated below.

The performance of different models is graphically represented as shown in Figure 6.7. The LSTM model has performed well when compared with the other models.

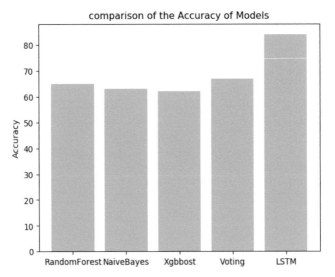

Figure 6.7   Performance of the classifiers.

6.7 Evaluation Parameters

Machine learning algorithms mainly use four available classification performance metrics [19], which are "binary classification, multi-class classification, multi-topic classification, and hierarchical classification" metrics. To
avoid uncertainty when learning performance metrics, the confusion matrix
is used to provide the best result. The accuracy of the classification model is
measured by counting the number of samples that are accurately identified
category, which is called "true positives (TP)." If the sample is identified but
not related to the desired category, it is a "true negative (TN)." If the sample
is not identified and is not associated with any category, it is a "false positive
(FP)." If they are not recognized as class samples, they are "false negatives
(FN)." These four values organize a confusion matrix. The confusion matrix
mainly describes the classifier's performance on a test dataset constructed
in the tabular form to identify the real value. Considering these definitions,
the following group of measures to evaluate the performance of the classifier
such as accuracy, precision, recall, and $F1$-score [17, 18, 20]. The measurement accuracy is defined in the following equation:

$$Accuracy = \frac{(TP + TN)}{(P + N)} \tag{6.3}$$

where TP is the number of true positives, TN is the number of true negatives,
and $P + N$ is the total population.

Precision is defined as the number of true positives divided by the total number of true positives and false positives [18]. This will tell how many retrieved elements are relevant to total positive populations, as given in the following equation:

$$Precision = \frac{TP}{TP + FP} \qquad (6.4)$$

Recall: Recall, also known as "sensitivity," is calculated by dividing the total number of true positives by the sum of true positives and false negatives [18]. This will tell about how many retrieved elements are relevant to the total population as given in the following equation:

$$Recall = \frac{TP}{TP + FN} \qquad (6.5)$$

*F*1-score: Usually, the two above metrics such as precision and recall are inversely propositional to each other. If precision value increases, the recall value decreases and vice versa. *F*1-score is "average of precision and recall. It takes both false positives and false negatives into consideration," which is given in eqn (6.6). It will be very helpful when there is skewness in class distribution.

$$F1 - Score = \frac{2 \times (Precision \times Recall)}{(Precision + Recall)} \qquad (6.6)$$

6.8 Conclusion and Future Work

Personalized medicine involves the systematic study of genetic mutation and other related information. Understanding the cancer tumor growth is a challenging task even when the advanced genetic analysis is adopted. Manually analyzing the genetic sequence will take much time. So the proposed method introduced a machine learning model to automate the process. The chapter combines both the LSTM and word embedding technique to enhance the performance. In addition, the chapter compared different word embedding with different classification models. The classification models like random forest, naive Bayes, XGBoost, and voting classifier were adapted. The proposed model integrates Word2Vec embedding and LSTM, which gives better results when compared to other existing classification models. The word embedding technique is mainly used to understand the semantic meaning of clinical annotations, which will enhance performance. It achieves 84% of accuracy.

Further, the proposed method can be extended to enhance the performance while introducing a genetic algorithm for feature selection.

References

[1] Mikolov, T., Chen, K., Corrado, G., & Dean, J. (2013). Efficient estimation of word representations in vector space. arXiv preprint arXiv:1301.3781.

[2] Mikolov, T., Sutskever, I., Chen, K., Corrado, G. S., & Dean, J. (2013). Distributed representations of words and phrases and their compositionality. In Advances in neural information processing systems (pp. 3111–3119).

[3] Le, Q., & Mikolov, T. (2014, June). Distributed representations of sentences and documents. In International conference on machine learning (pp. 1188–1196). PMLR.

[4] Kusner, M., Sun, Y., Kolkin, N., & Weinberger, K. (2015, June). From word embeddings to document distances. In International conference on machine learning (pp. 957–966). PMLR.

[5] Cohen, R., Aviram, I., Elhadad, M., & Elhadad, N. (2014). Redundancy-aware topic modeling for patient record notes. PloS one, 9(2), e87555.

[6] Marafino, B. J., Davies, J. M., Bardach, N. S., Dean, M. L., & Dudley, R. A. (2014). N-gram support vector machines for scalable procedure and diagnosis classification, with applications to clinical free text data from the intensive care unit. Journal of the American Medical Informatics Association, 21(5), 871–875.

[7] Wang, L., Chu, F., & Xie, W. (2007). Accurate cancer classification using expressions of very few genes. IEEE/ACM Transactions on computational biology and bioinformatics, 4(1), 40–53.

[8] Kim, S. B., Han, K. S., Rim, H. C., & Myaeng, S. H. (2006). Some effective techniques for naive bayes text classification. IEEE transactions on knowledge and data engineering, 18(11), 1457–1466.

[9] Baker, S., Korhonen, A. L., & Pyysalo, S. (2017). Cancer hallmark text classification using convolutional neural networks.

[10] Chiu, B., Crichton, G., Korhonen, A., & Pyysalo, S. (2016, August). How to train good word embeddings for biomedical NLP. In Proceedings of the 15th workshop on biomedical natural language processing (pp. 166-174).

[11] Pennington, J., Socher, R., & Manning, C. D. (2014, October). Glove: Global vectors for word representation. In Proceedings of the 2014 conference on empirical methods in natural language processing (EMNLP) (pp. 1532–1543).

[12] Maas, A., Daly, R. E., Pham, P. T., Huang, D., Ng, A. Y., & Potts, C. (2011, June). Learning word vectors for sentiment analysis. In Proceedings of the 49th annual meeting of the association for computational linguistics: Human language technologies (pp. 142–150).

[13] Vieira, J. P. A., & Moura, R. S. (2017, September). An analysis of convolutional neural networks for sentence classification. In 2017 XLIII Latin American Computer Conference (CLEI) (pp. 1–5). IEEE.

[14] Yaghoobzadeh, Y., & Schütze, H. (2016). Intrinsic subspace evaluation of word embedding representations. arXiv preprint arXiv:1606.07902.

[15] Baker, S., Korhonen, A. L., & Pyysalo, S. (2017). Cancer hallmark text classification using convolutional neural networks.

[16] Pennington, J., Socher, R., & Manning, C. D. (2014, October). Glove: Global vectors for word representation. In Proceedings of the 2014 conference on empirical methods in natural language processing (EMNLP) (pp. 1532–1543).

[17] Sundari, P. S., & Subaji, M. (2020). An improved hidden behavioral pattern mining approach to enhance the performance of recommendation system in a big data environment. Journal of King Saud University-Computer and Information Sciences.

[18] Novaković, J. D., Veljović, A., Ilić, S. S., Papić, Ž., & Milica, T. (2017). Evaluation of classification models in machine learning. Theory and Applications of Mathematics & Computer Science, 7(1), 39–46.

[19] Bost, R., Popa, R. A., Tu, S., & Goldwasser, S. (2015, February). Machine learning classification over encrypted data. In NDSS (Vol. 4324, p. 4325).

[20] Parwez MA, Abulaish M. Multi-label classification of microblogging texts using convolution neural network. IEEE Access. 2019 May 27;7:68678–91.

[21] Chiu B, Baker S. Word embeddings for biomedical natural language processing: survey. Language and Linguistics Compass. 2020 Dec;14(12):e12402.

[22] Gerevini AE, Lavelli A, Maffi A, Maroldi R, Minard AL, Serina I, Squassina G. Automatic classification of radiological reports for clinical care. Artificial intelligence in medicine. 2018 Sep 1;91:72–81.

[23] Baker S. Semantic text classification for cancer text mining (Doctoral dissertation, University of Cambridge).

[24] Parwez MA, Abulaish M. Multi-label classification of microblogging texts using convolution neural network. IEEE Access. 2019 May 27;7:68678–91.

[25] Jiang L, Sun X, Mercaldo F, Santone A. DECAB-LSTM: Deep Contextualized Attentional Bidirectional LSTM for cancer hallmark classification. Knowledge-Based Systems. 2020 Dec 27;210:106486.

[26] Abhishek P Iyer, J. Karthikeyan, MD. Rakibul Hasan Khan, P.M. Binu (2020) An analysis of Artificial Intelligence in Biometrics-The next level of security. Journal of Critical Reviews, 7 (1), 571–576.

[27] Pallavi Verma, Saksham Bhutani, S. Srividhya, J Karthikeyan,Chong Seng Tong (2020) Review of internet of things towards sustainable development in agriculture. Journal of Critical Reviews, 7 (3), 148–151.

[28] Karthikeyan, J., & Shiny, K. G. (2019). A technology integrated analysis of educational and ELT research process. Journal of Advanced Research in Dynamical and Control Systems, Vol.11, 9- Special Issue, 821–827.

7

Prediction of Covid-19 Disease using Machine-learning-based Models

Abhisek Omkar Prasad, Megha Singh, Pradumn Kumar Mishra, Shubhi Srivastava, Dibyani Banerjee, and Abhaya Kumar Sahoo

School of Computer Engineering, KIIT Deemed to be University, India
Email: 1929302@kiit.ac.in; 1928300@kiit.ac.in; 1929299@kiit.ac.in; 1928122@kiit.ac.in; 1905750@kiit.ac.in; abhayakumarsahoo2012@gmail.com

Abstract

According to Chinese health officials, almost 250 million people in China may have caught Covid-19 in the first 20 days of December. Due to the Covid-19 pandemic and its global spread, there is a significant impact on our health system and economy, causing many deaths and slowing down worldwide economic progress. The recent pandemic continues to challenge the health systems worldwide, including a life that realizes a massive increase in various medical resource demands and leads to a critical shortage of medical equipment. Therefore, physical and virtual analysis of day-to-day death, recovery cases, and new cases by accurately providing the training data are needed to predict threats before they are outspread. Machine learning algorithms in a real-life situation help the existing cases and predict the future instances of Covid-19. Providing accurate training data to the learning algorithm and mapping between the input and output class labels minimizes the prediction error. Polynomials are usually used in statistical analysis. Furthermore, using this statistical information, the prediction of upcoming cases is more straightforward using those same algorithms. These prediction models combine many features to predict the risk of infection being developed. With the help of prediction models, many areas can be strengthened beforehand to cut down risks and maintain the health of the citizens. Many predictions before the second wave of Covid-19 were realized to be accurate, and if we had worked on it, we would have decreased the

109

fatality rate in India. In particular, nine standard forecasting models, such as linear regression (LR), polynomial regression (PR), support vector machine (SVM), Holt's linear, Holt–Winters, autoregressive (AR), moving average (MA), seasonal autoregressive integrated moving average (SARIMA), and autoregressive combined moving average (ARIMA), are used to forecast the alarming factors of Covid-19. The models make three predictions: the number of new cases, deaths, and recoveries over the next 10 days. To identify the principal features of the dataset, we first grouped different types of cases as per the date and plotted the distribution of active and closed cases. We calculated various valuable stats like mortality and recovery rates, growth factor, and doubling rate. Our results show that the ARIMA model gives the best possible outcomes on the dataset we used with the most minor root mean squared error of 23.24, followed by the SARIMA model, which offers somewhat close results to the AR model. It provides a root mean square error (RMSE) of 25.37. Holt's linear model does not have any considerable difference with a root mean square error of 27.36. Holt's linear model has a value very close to the moving average (MA) model, which results in the root mean square of 27.43. This research, like others, is also not free from any shortcomings. We used the 2019 datasets, which missed some features due to which models like Facebook Prophet did not predict results up to the mark; so we excluded those results in our outcomes. Also, the python package for the Prophet is a little non-functional to work on massive Covid-19 datasets appropriately. The period is better, where there is a need for more robust features in the datasets to support our framework.

7.1 Introduction

In machine learning, artificial intelligence is a branch of computer science that enables machines to do their jobs more successfully by utilizing intelligent algorithms. Furthermore, machine learning helps to solve the real-world problems. Forecasting, stock market forecasts, and disease prognosis are some of the essential uses of machine learning. Machine learning techniques have been applied to predict various diseases such as coronary artery disease, breast cancer, and cardiovascular disease [9, 10, 21]. Lately, machine learning (ML) algorithms in real-life situations help the existing cases and predict the future instances of Covid-19 [31]. Moreover, ML algorithms are utilized to recognize small collective behavior and utilize real-time data to predict how the Covid-19 is likely to spread across society. Corona virus disease, which emerged in 2019, is the latest global pandemic to cause severe health threats and economic consequences [1, 22, 23]. China's Wuhan City experienced an outbreak of Covid-19 in December. Disease outbreaks swept through many countries in the early stages, with Japan, Thailand, and South Korea being the most affected. Viruses that cause severe acute respiratory syndrome (SARS), which was responsible for the disease, were designated as Covid-19 by the WHO [27]. Generally, the poor and underdeveloped are the actual sufferers of infectious diseases. The corona virus proved that even developed countries are not immune to infectious diseases by providing us with sufficient evidence supporting this belief. As part of our study, we intend to develop a Covid-19 forecasting system to contribute to the current humanitarian crisis [6, 7, 28]. In order to forecast the disease for the next 10 days, three essential variables are used: (a) total confirmed cases, (b) total death cases, and (c) total recovered cases.

The remaining parts of the chapter are organized as follows. Section 7.2 explains the literature survey. Covid-19 disease predictions are discussed in Section 7.3. Section 7.4 presents evaluation parameters used in models. Section 7.5 explains the analysis of experimental result using different models and Section 7.6 shows the conclusion with future work.

7.2 Literature Survey

Every person has been affected by Covid-19 in their own way. Covid-19 was given its name by the World Health Organization (WHO) on February 11, 2020 [26, 29]. It is a deadly virus as it multiplies rapidly and exponentially. It can spread from patient to patient via tiny droplets when they cough or exhale. Therefore, it is easy for other people to catch Covid-19 from touching these objects or surfaces. Due to this, maintaining social distance is imperative. Wearing masks, washing your hands properly, maintaining the hygiene by being self-sanitized, and staying in your home are all advised. Most common and severe symptoms include fever, dry cough, fatigue, chest pain, difficulty speaking or moving, shortness of breath, loss of speech, and shortness of breath [15–20, 30]. The forecasting system of Covid-19 uses nine standard forecasting models, such as Holt's linear model, linear regression, Holt–Winters model, polynomial regression, moving average model, SVM, AR model, ARIMA model, and SARIMA model [5].

In Holt's linear model, constant movement is observed, whether increasing or decreasing. It is mainly suitable for time series datasets that have trend components but lack seasonal details. Holt–Winters also describes the trends in time series and attempts to capture three aspects of the trends, namely a typical value, a slope over time, and the seasonality of a series. In linear regression, an independent variable is used to estimate the effect of the dependent variable. Therefore, the linking between independent and dependent variables as well as forecasting can be found using this method. The polynomial regression algorithm builds the association between independent variable and dependent variable. Using a linear function, SVM solves regression problems. The SVM maps inputs into an *n*-dimensional space called a feature space when dealing with non-linear regression issues. The moving average model is also a time average model based on the data provided of the previous time spots. It chiefly uses the errors mentioned above to estimate the current period values' better-period. If we need a linear combination of variables' past values, the AR model is used. The ARIMA model, which is among the most frequently used forecasting methods, is mainly used as a statistical estimation test polynomial for analyzing and forecasting time series data. The SARIMA model extends ARIMA; it integrates seasonal elements, allowing direct modeling of seasonal changes [8].

7.3 Different Models used in Covid-19 Disease Prediction

For the Covid-19 forecast, we used the following nine prediction models: Holt's linear model, Holt–Winters model, linear regression, polynomial regression, SVM prediction, moving average model (using AUTO ARIMA), AR model, ARIMA model, and SARIMA model [11, 12].

7.3.1 Holt's linear model

Holt's linear model comes into the category of linear, exponential smoothing models. It comes into the process when we require forecasting data with trends. There are three separate equations that work together to generate forecasts. The forecasts generated by this model show a constant movement, whether increasing or decreasing. Here in our prediction, time series forecasting uses Holt's linear trend method. Usually, it is suitable for time series datasets that have trend components but lack seasonal details. In this procedure, a single forecast equation and two smoothing equations are combined, i.e., level one and the trend one.

7.3.2 Holt–Winters method

Among the components of the Holt–Winters seasonal method are a forecast equation and three smoothing equations. One for the level l_t, one for the trend b_t, and one for the seasonal component s_t. Smoothing parameters consist of α, β^*, and γ. The parameter m represents seasonality. It is the number of seasons per year. This technique comprises four methods such as Holt–Winters

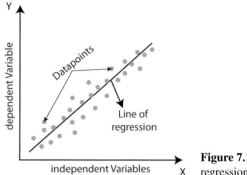

Figure 7.1 Graph that represents line of regression.

ES, Holt's ES, exponential smoothing, and weighted averaging. As long as seasonal variations change proportionally to levels in a series, the multiplicative method is preferred. However, we used both seasonally and trend-based smoothing equations [14].

7.3.3 Linear regression

This is a statistical method for analyzing predictions using linear regression. Variables that are continuous/actual or numeric are predicted by linear regression. Based on independent features of a class, a regression model establishes the target class. This method will help you determine how independent variables impact a dependent variable. In order to determine whether an independent variable has an impact on a dependent variable, it is used. Therefore, the linking between independent and dependent variables as well as forecasting can be found using this method, which is represented in Figure 7.1 [13].

Linear regression involves two factors (x, y). Regression variable y is sometimes called a regress band, dependent variable, or explained variable. A predictor variable x is also referred to as a regression variable, an independent variable, or an explanation variable [25]. Mathematically, a linear regression can be expressed as follows:

$$z = b0 + b1k + \varepsilon, \tag{7.1}$$

where z represents target variable or dependent variable, k denotes independent variable or predictor variable, $b0$ denotes an intercept of a line, $b1$ represents scale factor for the input value, and ε denotes random error. Train the machine learning model using the linear regression concept. An input training dataset is represented by k, and a set of class labels is represented by z. By finding the best values for $b0$ (intercept) and $b1$ (coefficient), the machine learning algorithm finds the best regression line. A minimum error should exist between the predicted and actual values to get the best-fit line with the minor error. Therefore, we need to calculate the best deals of $b0$ and $b1$ using the cost function. To calculate the cost function, we use mean squared error (MSE). The average of squared errors between predicted and actual values is used.

7.3.4 Polynomial regression

A polynomial regression is a method of regression in which the relationship between the independent variables (x) and the dependent variables (y) is expressed as a polynomial. The conditional mean of y is represented by $E(y, x)$ and can be fitted to the value of x, not via linear relationships [13]. An estimation test is based on statistical data. For unknown parameters assumed in the experiment, polynomial regression fits a non-linear model since the regression function $E(x, y)$ is linear. Multiple linear regressions are therefore considered polynomial regressions. Based on an independent variable x, we can predict a dependent variable y by analyzing regression analysis. A general polynomial regression model is expressed in eqn (7.2) when we square an nth-degree polynomial, thereby obtaining the expected value of y:

$$y = \beta 0 + \beta 1 x + \beta 2 x^2 + \beta 3 x^3 + \cdots + \beta X^n + c. \tag{7.2}$$

These models are convenient as far as estimation is concerned because the regression function, based on the unknown variables $\beta 0$, $\beta 1$, etc., is linear [25].

7.3.5 Support vector machine

Support vector machines (SVMs) are the dominant algorithms in supervised machine learning. An n-dimensional hyper-plane is found using a support vector machine algorithm that distinctly classifies the data points. As SVM regression is a non-parametric technique, it relies on mathematical functions. An SVM algorithm alters the input data space with a kernel set based on the requirements. Kernel tricks are used by SVM for non-linearly separable data to make it linearly separable. The better period on Cover's theorem states that, given a set of non-linearly separable data, it is likely to be transformed into a linearly separable one through some non-linear transformation. Tricks help project data points to the higher-dimensional space by which they become relatively more easily detachable in higher-dimensional areas [4]. Regression problems are solved by using a linear function.

7.3.6 Moving average model

An example of a time average model is the moving average model that is based on the data provided of the previous time spots. It is entirely based on past errors in the series, known as error lags. Defining it technically, the MA model is a filter applied to the random signal (white noise) with some other interpretation. The filter is a finite impulse response filter whose impulse response settles to zero in limited time, i.e., finite duration. To summarize each, the MA model chiefly uses the previous errors to estimate the current period values' better-period. In addition, the moving average model is theoretically equivalent to a linear regression based on the current and observed error terms of the previous period. We assume that the error terms are also mutually independent and have the same normal distribution.

7.3.7 Autoregressive model

The autoregressive model comes into the category of the linear combination of predictors. Whenever a linear combination of previously recorded values of a variable is needed, it enters the process. Using the autoregressive model, the output variable varies linearly as a function of its previous values and on a stochastic term (unpredictable); therefore, the model can be represented as latent difference equation. Here, in our prediction, we have done time series forecasting with the autoregressive model method. Usually, it is suitable for time series datasets that have trend components but lack seasonal features. Eqn (7.3) presents a model of autoregression of order p:

$$Y_t = c + \phi_1 y_{t-1} + \phi_2 y_{t-2} + \cdots + \phi_p y_{t-p} + e_t. \tag{7.3}$$

Despite their flexibility, autoregressive models are capable of handling many types of time series patterns. By changing $\phi_1, \ldots, \phi_p$, different time series patterns can be derived. Changing e_t will only change its scale, not its patterns.

7.3.8 ARIMA

Future values are linear functions of past values in ARIMA models; so future values are linear functions of past values. The ARIMA model is among the most commonly used forecasting methods for univariate time series data. Even though the technique can handle data with trends, it is unable to handle time series with seasonal components. An ARIMA model (p, d, q) can be the number of moving averages and the number of differences is the number of autoregressive terms.

7.3.8.1 Seasonal ARIMA model

In forecasting univariate time series data, the ARIMA model is the most commonly used. Even though the technique can handle their data with trends, it cannot manage seasonal time series. Seasonal changes can be directly modeled with SARIMA, an extension of ARIMA that incorporates the seasonal element. SARIMA, or seasonal ARIMA, extends ARIMA to account for time series with a seasonal component. Three new hyper-parameters, including auto regression (AR), differentiation (*I*), and moving average (MA), are used to specify the seasonality of the series, along with a parameter for the period of seasonality. By incorporating seasonal terms into the existing ARIMA models, we can create a seasonal ARIMA model.

7.4 Evaluation Parameters used in Models

For each of the learning models, we evaluate their accuracy based on three metrics: mean square error (MSE), mean absolute error (MAE), and root mean square error (RMSE).

7.4.1 Root mean square error

The root mean square error is the standard deviation of the prediction errors. The residuals of a prediction are also known as prediction errors. A residual is a distance between a line that fits the data and the actual data points. The best-fit line can be determined by looking at how the data points cluster around it; we can get an idea of how it presents the essential data [24].

$$\text{RMSE} = \sqrt{\frac{\sum_{i=1}^{N}(O-P)^2}{N}}, \tag{7.4}$$

where N represents the number of data points, O represents the observed value, and P represents the predicted value.

7.4.2 Mean square error

As an estimation error, the mean square error is the average of all squares resulting from the difference between the actual value and the estimated value. We find the square of the data points based on the distance of the data points from the regression line. By squaring values, all negative values are eliminated and differences that are the most significant are emphasized.

$$\text{MSE} = \sqrt{\frac{\sum_{i=1}^{N}(O-P)^2}{N}}, \tag{7.5}$$

where N represents the number of data points, O represents the observed value, and P represents the predicted value.

7.4.3 Mean absolute error

In regression models, the mean absolute error is used. It is the mean of the absolute values of errors in a set of predictions from the model. Therefore, the MAE of a model concerning a test set can be defined as the mean of the fundamental values of the individual prediction errors on all instances in the testing set. A prediction error, for example, is a difference between the actual and the predicted value [24].

$$\text{MAE} = \frac{\sum_{i=1}^{N}|(P-T)|}{N}, \tag{7.6}$$

where a total number of data points is N, a predicted value is P, and a real value is T.

7.5 Experimental Result Analysis

This study employs machine learning to estimate the number of cases affected by Covid-19 using a forecasting tool. The study used data from the daily reports on Covid-19 reported worldwide about the number of cases newly identified, the number of recoveries, and the number of deaths caused by Covid-19. The global situation has become increasingly troubling day after day, more and more cases are being confirmed, and deaths are becoming more frequent. Covid-19 could potentially affect people worldwide but has not been studied yet in various countries. Researchers hope to estimate how many new infections and deaths will occur over the next 10 days and how many people will recover. Nine machine learning models, such as Holt's linear model, moving average model (using AUTO ARIMA), linear regression, polynomial regression, SVM prediction, Holt–Winters model, AR model, ARIMA model, and SARIMA model, have been utilized to count the number of deaths, new infections, and recoveries.

Table 7.1 Models performance for predicting death rates in the future.

Model used	MSE	MAE	RMSE
LR	260,639,650,567.8839	509,202.63	510,528.79
Polynomial regression	93,282,774,399.11284	253,618.2916	305,422.2886
SVM	2,584,525,668,527.067	1,579,450.6289	1,607,646.0022
AR	2.6139	1615.3588	161,678,100.9321
Holt's linear	10,540,774.6358	2868.83	3246.6559
Holt–Winters	148,489,136.0289	9220.9396	12,185.6118
MA	173,765,170.8795	9890.8186	13,182.0017
ARIMA	475,925,665.2412	132.2584	2185.7205
SARIMA	112,783,259.5446	7376.4914	10,619.9463

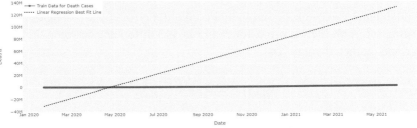

Figure 7.2 Death cases forecasting graph using linear regression.

7.5.1 Future forecasting of death rates

According to the results obtained, the ARIMA model performed the best with a minor root mean square when predicting the death rate. In comparison, linear regression model performed poorly in all the metrics. We received the minor mean squared error and somewhat satisfactory results in other metrics of the AR model. Further, Holt–Winters and moving average models outperformed others with fewer root mean square error metrics and mean absolute errors. The ARIMA model has the minor root mean squared error. Results of the prediction of all the models are displayed in Table 7.1.

The prediction graphs of all nine models are shown in Figure 7.2–7.10.

7.5.2 Future forecasting of confirmed cases

A growing number of cases of Covid-19 are confirmed every day. In this study, we used forecasting results shown in Table 7.2. The AR model and the SARIMA model performed best in future forecasting of confirmed cases. Also, MA and ARIMA models performed relatively better and are almost

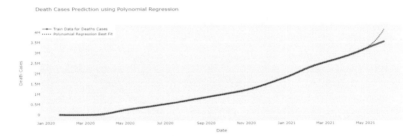

Figure 7.3   Death cases prediction graph using polynomial regression.

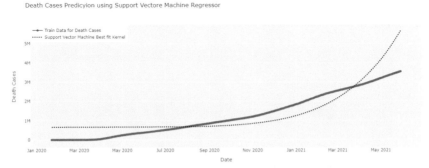

Figure 7.4   Death cases prediction graph using SVM.

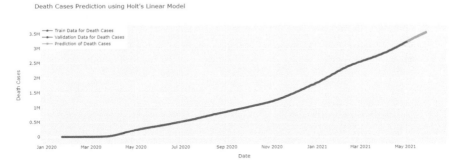

Figure 7.5   Death cases prediction graph using the Holt's linear model.

equal. But the Holt–Winters and the Holt's linear model showed comparatively less performance compared to the expected result as shown in the graph. According to each evaluation metric, LR did not perform well. In contrast, we received satisfactory results for MSE using SVM, and the rest of the evaluating metrics of SVM shows poor results.

The prediction graphs of all nine models are shown in Figures 7.11–7.19.

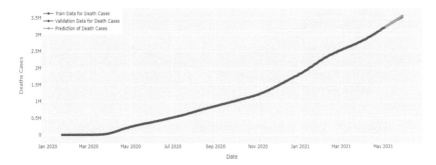

Figure 7.6   Death cases forecasting graph using the Holt–Winters model.

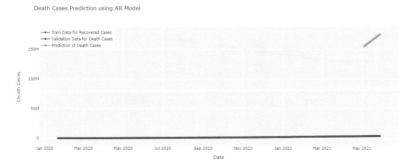

Figure 7.7   Death cases prediction graph using the AR model.

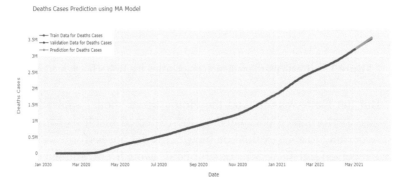

Figure 7.8   Death cases prediction graph using the MA model.

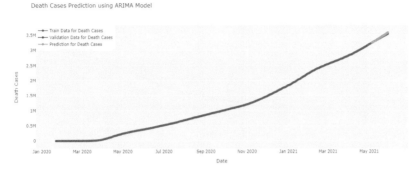

Figure 7.9 Death cases prediction graph using the ARIMA model.

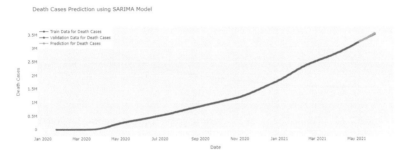

Figure 7.10 Death cases prediction graph using the SARIMA model.

Table 7.2 Models performance for predicting confirmed cases in the future.

Model used	MSE	MAE	RMSE
LR	7,763,724,965,004,040.00	88,061,662.55	88,112,002.34
Polynomial regression	748,731,493,306,401.81	23,242,128.97	27,362,958.42
SVM	2.5063797835361292	158,265,585.13	158,315,500.93
AR	5,527,034,041,537.31	1733.51	2,350,964.49
Holt's linear	2,876,795,212,473.51	1,205,603.52	1,696,111.79
Holt–Winters	5,999,012,879,456.42	1,870,166.76	2,449,288.24
MA	8,418,579,969,402.77	2,152,410.34	2,901,478.93
ARIMA	9,986,448,460,476.88	1540.25	3,160,134.25
SARIMA	5,559,283,753,365.25	1,731,905.12	2,357,813.34

7.5.3 Future forecasting of recovery rate

Among all the models, the ARIMA performs better in recovery rate and future forecasting. Considering the availability of time series data, all other models have poor results; ARIMA performs the best, followed by MA, AR, and

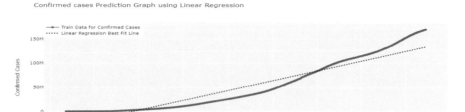

Figure 7.11   Confirmed cases prediction graph using linear regression.

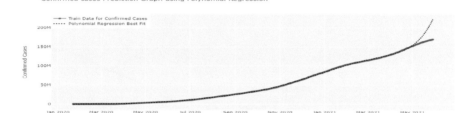

Figure 7.12   Confirmed cases forecasting graph using polynomial regression.

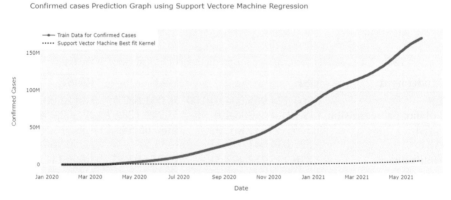

Figure 7.13   Confirmed cases prediction graph using SVM.

SARIMA models. Below are predictions for the next few days. In Table 7.3, we see the results of the various learning models:

Nevertheless, the ARIMA prediction closely tracks the predicted trends compared with the current state of the recovery. As predicted by our models, the recovery rate will be enhanced in the near future, and the graph shows the growth rate will be expanded while the death rate will be reduced. Prediction

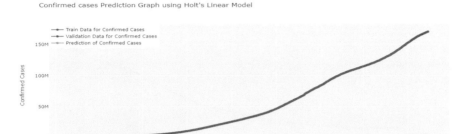

Figure 7.14 Confirmed cases prediction graph using the Holt's linear model.

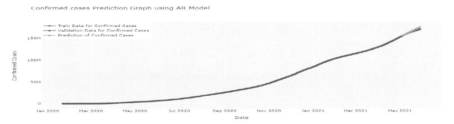

Figure 7.15 Confirmed cases prediction graph using the Holt–Winters model.

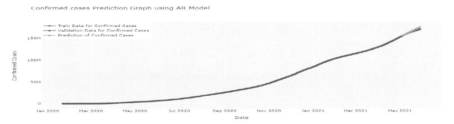

Figure 7.16 Confirmed cases prediction graph using the AR model.

plots of recovery rates according to the different models are represented in Figures 7.20–7.28.

7.6 Conclusion

Taking preventive measures is essential for limiting the spread of Covid-19. The government is taking different measures to prevent the current situation.

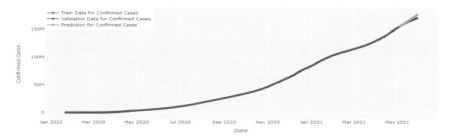

Figure 7.17 Confirmed cases prediction graph using the MA model.

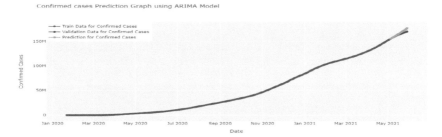

Figure 7.18 Confirmed cases forecasting graph using the ARIMA model.

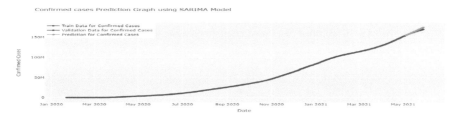

Figure 7.19 Confirmed cases forecasting graph using the SARIMA model.

Table 7.3 Model performance for predicting the recovery rate in the future.

Model used	MSE	MAE	RMSE
LR	610,646,062,724,160.8	24,485.06	24,711.76
Polynomial regression	45,402,037,557,809.54	52,858.37	67,381.40
SVM	229,188,868,590,859.0	14,517.18	15,138.05
AR	65,315,739.78	6530.07	6531.78
Holt's linear	7,409,386,501.91	16,260.00	27,220.88
Holt–Winters	641,256,728.24	45,176.70	80,078.09
MA	405,405,694,677.06	5875.90	63,671.76
ARIMA	1,526,842,334,364.99	1069.09	12,356.61
SARIMA	7,464,582,539,343.50	24,388.91	27,321.82

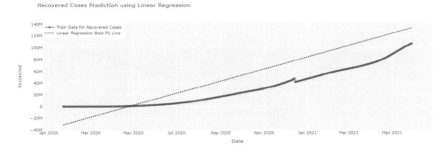

Figure 7.20 Recovered cases prediction graph using linear regression.

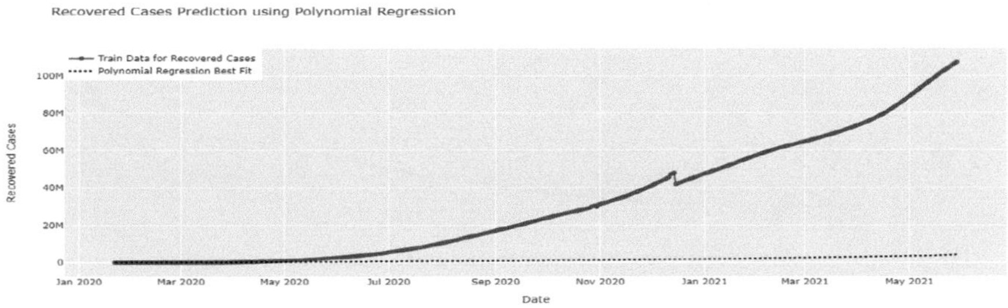

Figure 7.21 Recovered cases prediction graph using polynomial regression.

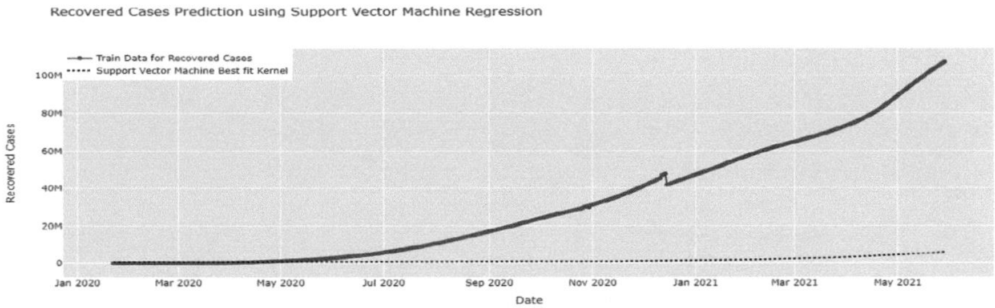

Figure 7.22 Recovered cases prediction graph using SVM.

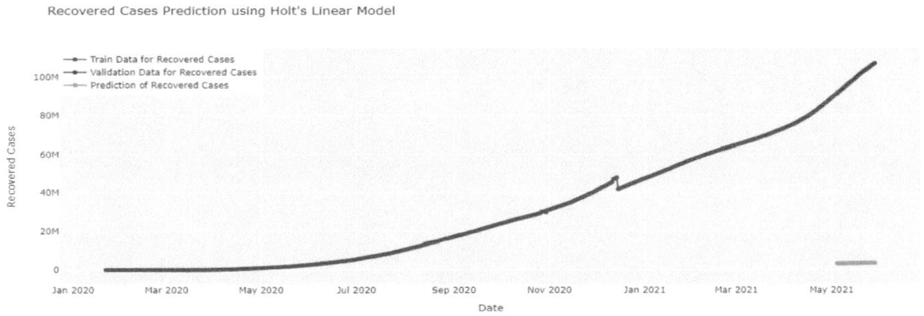

Figure 7.23 Recovered cases prediction graph using the Holt's linear model.

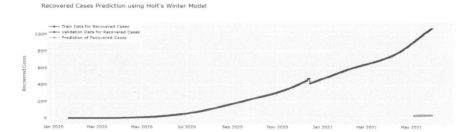

Figure 7.24   Recovered cases prediction graph using the Holt–Winters model.

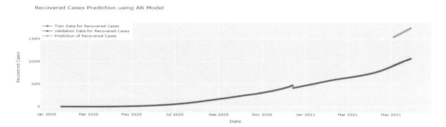

Figure 7.25   Recovered cases prediction graph using the AR model.

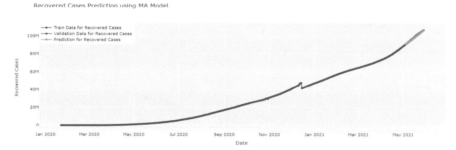

Figure 7.26   Recovered cases prediction graph using the MA model.

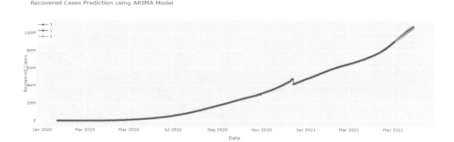

Figure 7.27   Recovered cases prediction graph using the ARIMA model.

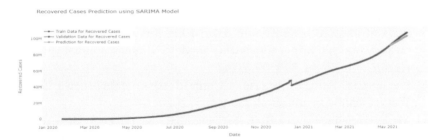

Figure 7.28 Recovered cases prediction graph using the SARIMA model.

ML-based prediction systems have been found to be most effective in predicting global Covid-19 outbreak trends in this study. Using various machine learning algorithms, the system analyzes the dataset that consists of actual past data day-by-day and predicts the upcoming days. Given the nature and size of datasets in this forecasting domain, the ARIMA model on average performs the best, followed by the SARIMA model, which offers similar results to the ARIMA model. The Holt's linear model has a very close value to the moving average (MA) model. Despite the variability in the dataset values, SVM produces poor results regardless of the scenario. As a result, the model predictions are correct for the current scenario, which can help us understand what will happen in the next few years.

References

[1] B. Tang, X. Wang, Q. Li, N. L. Bragazzi, S. Tang, Y. Xiao, and J. Wu, "Estimation of the transmission risk of the 2019-ncov and its implication for public health interventions," Journal of clinical medicine, vol. 9, no. 2, p. 462, 2020.

[2] W. Wei, J. Jiang, H. Liang, L. Gao, B. Liang, J. Huang, N. Zang, Y. Liao, J. Yu, J. Lai et al., "Application of a combined model with autoregressive integrated moving average (arima) and generalized regression neural network (grnn) in forecasting hepatitis incidence in heng county, china," PloS one, vol. 11, no. 6, p. e0156768, 2016.

[3] H. Tandon, P. Ranjan, T. Chakraborty, and V. Suhag, "Coronavirus (covid-19): Arima based time-series analysis to forecast near future," arXiv preprint arXiv:2004.07859, 2020.

[4] W.S. Noble, Support vector machine applications in computational biology. in Kernel Methods in Computational Biology (eds. Schoelkopf, B., Tsuda, K. & Vert, J.-P.) 71–92 (MIT Press, Cambridge, MA, 2004).

[5] L. Moftakhar, S. Mozhgan, and M. S. Safe, "Exponentially increasing trend of infected patients with covid-19 in iran: A comparison of neural network and arima forecasting models," Iranian Journal of Public Health, vol. 49, pp. 92–100, 2020.

[6] Pneumonia of unknown cause — China: disease outbreak news. Geneva: World Health Organization, January 5, 2020

[7] M. L. Holshue, C. DeBolt, S. Lindquist, K. H. Lofy, J. Wiesman, H. Bruce, & S. K. Pillai(2020). First case of 2019 novel coronavirus in the United States. New England Journal of Medicine.

[8] Adam D. Bull. Convergence rates of efficient global optimization algorithms. Journal of Machine Learning Research, (3-4):2879–2904, 2011.

[9] Coronavirus disease 2019 (COVID-19): situation report — 36. Geneva: World Health Organization, February 25, 2020

[10] WHO R&D blueprint: informal consultation on prioritization of candidate therapeutic agents for use in novel coronavirus 2019 infection. Geneva: World Health Organization, January 24, 2020

[11] S. Makridakis, E. Spiliotis, and V. Assimakopoulos, "Statistical and machine learning forecasting methods: Concerns and ways forward," PloS one, vol. 13, no. 3, 2018.

[12] G. Bontempi, S. B. Taieb, and Y.-A. Le Borgne, "Machine learning strategies for time series forecasting," in European business intelligence summer school. Springer, 2012, pp. 62–77.

[13] F. E. Harrell Jr, K. L. Lee, D. B. Matchar, and T. A. Reichert, "Regression models for prognostic prediction: advantages, problems, and suggested solutions." Cancer treatment reports, vol. 69, no. 10, pp. 1071–1077, 1985.

[14] C. Chatfield, M. Yar, Holt-Wintersforecasting: some practicalissues, The Statistician, Vol. 37, 1988, pp. 129-140.

[15] K. M. Anderson, P. M. Odell, P. W. Wilson, and W. B. Kannel, "Cardiovascular disease risk profiles," American heart journal, vol. 121, no. 1, pp. 293–298, 1991.

[16] X. Chen et al. Restoration of leukomonocyte counts is associated with viral clearance in COVID-19 hospitalized patients. Preprint at medRxiv

[17] J. Zhao et al. Antibody responses to SARS-CoV-2 in patients of novel coronavirus disease 2019. Clin. Infect. Dis.

[18] C. Shen et al. Treatment of 5 critically ill patients with COVID-19 with convalescent plasma. JAMA

[19] Y. Shi et al. Immunopathological characteristics of corona virus disease 2019 cases in Guangzhou, China. Preprint at medRxiv

[20] L. Runfeng et al. Lianhuaqingwen exerts anti-viral and anti-inflammatory activity against novel coronavirus (SARS-CoV-2). Pharmacol. Res.

[21] Andrew Saxe, Pang Wei Koh, Zhenghao Chen, Maneesh Bhand, Bipin Suresh, and Andrew Ng. On random weights and unsupervised feature learning. In Proceedings of the 28th International Conference on Machine Learning, 2011.

[22] W. J. Guan, Z. Y. Ni, Y. Hu, W. H. Liang, C. Q. Ou, J. X. He, & N. S. Zhong (2020). Clinical characteristics of 2019 novel coronavirus infection in China. MedRxiv.

[23] B. N. Iyke, (2020). Economic Policy uncertainty in times of COVID-19 pandemic. Asian Economics Letters 1:2. doi:10.46557/001c.17665.

[24] C. Willmott, and K. Matsuura: Advantages of the Mean Absolute Error (MAE) over the Root Mean Square Error (RMSE) in assessing average model performance, Clim. Res., 30, 79– 82, 2005. 1526, 1527, 1528, 1529

[25] R.Tibshirani, "Regression shrinkage and selection via the lasso," Journal of the Royal Statistical Society: Series B (Methodological), vol. 58, no. 1, pp. 267–288, 1996.

[26] Coronavirus 2019-nCoV, CSSE. Coronavirus 2019-nCoV Global Cases by Johns Hopkins CSSE. (Available from:https://gisanddata.maps.arcgis.com/apps/opsdashboard/index.html#/bda7594740fd40299423467b48e9ecf6)

[27] W Guan, Z Ni, H Yu. Clinical characteristics of 2019 novel coronavirus infection in China. medRxiv preprint posted online on Feb. 9, 2020; 10.1101/2020.02.06.20020974.

[28] M Wang, R Cao, L Zhang. Remdesivir and chloroquine effectively inhibit the recently emerged novel coronavirus (2019-nCoV) in vitro. Cell Res 2020. 10.1038/s41422-020-0282-0

[29] P Zhou, XL Yang, XG Wang. A pneumonia outbreak associated with a new coronavirus of probable bat origin. Nature 2020. 10.1038/s41586-020-2012-7

[30] Q Li, X Guan, P Wu. Early transmission dynamics in Wuhan, China, of novel coronavirus-infected pneumonia. N Engl J Med 2020. 10.1056/NEJMoa2001316

[31] A. K. Sahoo, C. Pradhan, & H. Das (2020). Performance evaluation of different machine learning methods and deep-learning based convolutional neural network for health decision making. In Nature inspired computing for data science (pp. 201-212). Springer, Cham.

8

Intelligent Retrieval Algorithm using Electronic Health Records for Healthcare

S. Nagarjuna Chary[1], N. Vinoth[2], and Kiran Chakravarthula[3]

[1]Department of Electronics & Instrumentation Engineering, VNR Vignana Jyothi Institute of Engineering and Technology, India.
[2]Department of Instrumentation, MIT Campus, Anna University, India.
[3]Department of Electronics & Instrumentation Engineering, VNR Vignana Jyothi Institute of Engineering and Technology, India.
Email: nagarjunachary_s@vnrvjiet.in; vinothbalaji@rediffmail.com; kiran_c@vnrvjiet.in

Abstract

The advent of artificial intelligence (AI) and machine learning (ML) in healthcare has been a major value addition and enrichment of the potential of what can be achieved using electronic health records (EHR) including radiology images, laboratory test results, time series data (such as electrocardiogram (ECG), electroencephalogram (EEG), etc.), as well as data collected from sensors like photoplethysmogram (PPG) and pulse oximeter (SpO_2). EHR can be more efficiently mined for different healthcare applications including but not limited to prognosis, diagnosis, and treatment. The machine learning technology in healthcare systems is most effective. There are many areas for the study in biomedical engineering, such as studies on genomics and proteins, biomedical imaging, brain–body–machine interface, and healthcare management based on publicly available data, i.e., EHR. This chapter presents a systematic approach on the use of machine learning techniques on large medical data (EHR) for classification, evaluation, and to improve decision making in healthcare system for the betterment of humanity. It may be noted that significant work has been done on wearable devices recently. PPG is widely used for various applications. This chapter discusses the use of PPG in biomedical engineering and healthcare applications. One can also investigate the study on data fusion, i.e., heart-rate detection from fusing various physiological parameters (such as those measured using ECG, PPG, and EEG, among others). Working with fused data is challenging, as sampling rate of each sensor varies, and the signal normalization and resampling are part of pre-processing. Machine learning can help address the complexity of prediction in healthcare analytics, whether it is classification or regression, as is evident from wide research in the area. Some of the optimal techniques are also discussed in the chapter.

8.1 Introduction

It is impossible to imagine healthcare solutions today without the advantages brought in by the introduction of digital technologies and computers. Computer processing of data has transformed the way healthcare is provided today. Modern healthcare, rightly, places patient safety at the highest priority. The problem arises when a patient is to be diagnosed about disorder [1] and conventional methods consume long wait times for patient diagnosis [2] and the reports thereof contain vast amounts of data, which require long processing time [3], resulting in increasingly extensive databases that cannot always be efficiently interpreted by clinicians or doctors. The application of artificial intelligence (AI) and machine learning (ML) in healthcare and medicine, especially in diagnosing disorders, creates avenues that can potentially outperform human interpretation of data [4]. Modern medical imaging and signal processing provide various features and attributes pertaining to the specific patient and derived from distinct types of analyses. The data in healthcare and clinical setting today effortlessly captures structured data regarding all aspects of care including medication, diagnosis, laboratory test results, and medical imaging data [5]. The patient data collected in digital form are typically referred to as electronic health records (EHRs) [6–8] or electronic medical records (EMRs) [8–11], which store medical records via computers. The EHRs are primarily used by physicians, clinicians, and researchers. The advent of ML improves the capability of not only classifying but also predicting abnormalities based on data analytics of the EMR [12]. Suitable pre-processing enables selecting appropriate features of interest in the EMR while deep learning (DL) [11, 13, 14] can improve accuracy of the predictions [15].

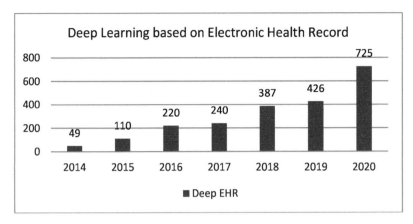

Figure 8.1 Year-wise publication in deep EHR. Data from Google scholar till September 2020.

8.2 EHR Datasets

An electronic health record (EHR) may be formally defined as a digital record of health information electronically stored as a systematized collection of individual population's medical information, which can be created, grouped, managed, and referred by authorized clinicians in healthcare organizations. A lot of advanced countries now routinely employ EHR, with every physician using EHR in Australia, closely followed by those in Netherlands, New Zealand, Israel, and Germany [9]. Canada and the United States had merely 25% of EHR before the Health Information Technology for Economic and Clinical Health (HITECH) [16] Act was introduced in 2009 with a $27 billion federal investment in developing and using technology for healthcare [17]. The proportion of hospitals using at least a basic EHR system has consequently risen up to over 80% while the research publications pertaining to deep learning and EHR has been dramatically increasing about 10 times in the last seven years as shown in Figure 8.1.

EHR systems log and store data related to each patient, which may include demographic details, diagnostic reports, details of laboratory tests and results, doctor prescriptions, radiological images, and clinical transcripts [17]. Through data analytics and machine learning, EHR can help doctors, hospitals, clinics, and other users only as may be authorized (Figure 8.2).

The processing of EHR data through analytics and machine learning is not simplistic because each patient's data is from multiple sources and is fundamentally heterogeneous. Machine learning on large datasets is facilitated through representation learning [18, 19], which refers to the portrayal

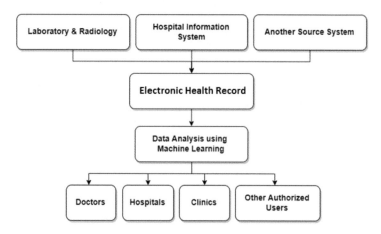

Figure 8.2 EHR process hierarchy.

of blended heterogeneous data using smaller elements for classifiers or prediction models that can help retrieve useful data. EMR-based profound learning [20] adds more layers of complexity owing to the need to protect patient's privacy and constrained data management following the patient's consent to share the data through certain pathways only [21].

8.2.1 EHR repositories

There are many repositories with huge and wide varieties of data, which are publicly available for research and study purpose; most of the datasets are provided as a service to the ML community. Some of the best dataset repositories for ML are as follows:

1. Google dataset search – https://datasetsearch.research.google.com/

2. Amazon datasets – https://registry.opendata.aws/

3. Microsoft datasets – https://azure.microsoft.com/en-in/services/open-datasets/

4. UCI machine learning repository – https://archive.ics.uci.edu/ml/index.php

5. Kaggle – https://www.kaggle.com/datasets

6. VisualData – https://visualdata.io/discovery

7. CMU libraries – https://guides.library.cmu.edu/az.php

The above-mentioned repositories ensure high quality and free dataset for study purpose.

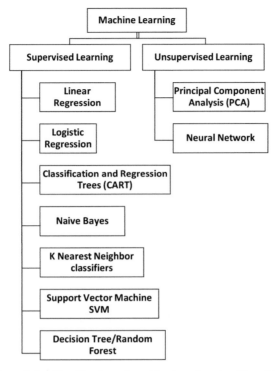

Figure 8.3 Classification of machine learning algorithms [44].

8.3 Machine Learning Algorithm

Scientists today concur that intelligent behavior of machines is founded in the machine's ability to learn. Artificial intelligence research cantered in ML is developing very rapidly, based on its relevance and affinity to man-made consciousness [41].

Machine learning methods can be broadly classified into two major groups, namely supervised and unsupervised learning. Supervised learning [35, 42] makes use of labeled data to learn, where the mapping function may be seen to be of the form $Y = f(X)$, indicating that the output is generated accurately for a given, new input data. On the other hand, unsupervised learning [18, 21, 43] uses data that are neither classified nor labeled. The ML algorithm must be able to act on these data without any guidance. Figure 8.3 [44] shows some of the common ML algorithms that can be applied to any practical data problem.

Table 8.1 Data mining techniques and algorithms.

Algorithm	Supervised	Unsupervised	Technique
Naive Bayes [45, 46]	Yes	–	Classification
Linear regression [47, 48]	Yes	–	Regression
ANN [49–51]	–	Yes	NN
K-means [52, 53]	–	Yes	Clustering
Quadratic discriminant analysis [35, 54–56]	Yes	Yes	Dimensionality reduction
SVM [57, 58]	Yes	–	Classification regression
k Nearest neighbor [59–61]	–	Yes	Clustering
Decision tree [62, 63]	Yes	–	Tree

Table 8.2 Important attributes for heart disease prediction [35].

Diseases	Causes
Coronary heart disease (CHD) or Coronary artery disease (CAD)	Tobacco use
Cerebrovascular disease (stroke)	Physical inactivity
Congenital heart disease	Harmful use of alcohol
Peripheral artery disease	Diabetes, smoking, obesity, and high BP
Inflammatory heart disease	Infectious agents, viruses, bacteria, fungi, or parasites
Rheumatic heart disease	Damage of heart valve, *Streptococcus* bacteria
Hypertensive heart disease	Unhealthy diet

8.3.1 Data mining techniques and algorithms

The major data mining methods and algorithm and its classification are presented in Table 8.1.

8.3.2 Machine learning algorithms using EHR for cardiac disease prediction

Electrocardiogram (ECG) is considered one of the most reliable and low-cost diagnostic tools to evaluate cardiac arrhythmia. Many research articles published earlier based on ECG for classification and prediction use conventional signal processing techniques Table 8.3. Now, with the advent of machine learning using EMR in healthcare, the problem of manual diagnosis of arrhythmia has been overcome. The machine learning approach for detection of ECG arrhythmia beats conventional approach of using the discrete wavelet transform (DWT) features [77]. Table 8.2 lists some of the most popular cardiac diseases and the corresponding causes.

Table 8.3 List of research papers on cardiovascular health study.

Domain	Author year	Disease	Method	Approach	Performance
Cardiovascular Health Study	Assodiky *et al.* [78]	Arrhythmia detection	DL and ANN	Particle swam optimization (PSO) and genetic algorithm (GA) for feature selection	DL is 76.51% with PSO as feature selection and 74.44 % with GA as feature selection
	Desai *et al.* [2]	Arrhythmia classification	DWT, ICA, and SVM	Five modules of cardiac arrhythmias are detected with good module specific accuracy	Extraction of hidden complex features with high discrimination
	Redmond *et al.* [42]	ECG signal quality and systemic vascular resistance	Supervised classification methodologies	Classification and discovery of health-related information from bio signals	ECG signal validation and estimating SVR have been provided
	Poplin *et al.* [79]	Cardiovascular risk factors	Deep-learning models	Using retinal fundus images	Predicted age *vs.* actual age and have a linear relationship
	Attila Reiss *et al.* [84]	Large-scale heart-rate estimation with CNN	Convolution neural network	PPG-based heart-rate estimation	CNN-based heart-rate prediction has outperformed

The majority of cardiovascular arrhythmia or diseases can be prevented, but demise keeps on ascending because of untimely treatment due to misdiagnosis. Here are some of the predictions done by researchers using ANN and DL methods, which outperform the results.

8.4 Machine Learning and Wearable Devices

In the recent years, wearable devices or simple wearables are becoming immensely popular and are finding applications in smartphones, smart watches, or even in apparels. The major advantage is that the typically MEMS-based sensor is small, consumes less power, and fits easily in any battery-powered device. These devices are aimed to sense various physiological activities of the subject and then produce meaningful data for further processing.

There is significant research and development going on in wearables for more accurate and reliable use. Some of the applications of biomedical wearable are real-time monitoring of heart functioning, heart-rate estimation, SpO_2 measurement, blood pressure monitoring, gait analysis, and diabetic monitoring.

Photoplethysmography is a simple, non-invasive, low-cost, and optical device that is primarily used to detect blood volume changes from the surface of skin. This optical device provides valuable information related to cardiovascular system. Figure 8.4 shows how a PPG device is attached to the skin in either transmission mode or in reflectance mode (A) and the working of the same when light is incident on the skin surface (B). When the infrared light passes through skin tissues, the blood absorbs the light strongly, thereby changing the intensity of light with respect to the blood flow. The voltage thus obtained from a PPG sensor is proportional to change in its reflected light intensity. Some of the major applications of PPG in biomedical wearables are blood oxygen saturation, heart-rate monitoring, blood pressure, respiration, and orthostasis.

The periodicity of the PPG signal matches the heartbeat and thus may be utilized to calculate heart rate (HR); HR knowledge aids in fitness tracking and can be used as a cardiovascular disease monitoring intervention mechanism on a regular basis. The PPG signals obtained from the wrist-worn device are sensitive to motion artifacts, due to movement of sensor module on the skin and/or sensor deformation due to long-term use. Thus, it affects the quality of signal that is extracted from the sensor. A significant work has been done on this area to remove the motion artifacts from the signal and estimate the heart rate [84]. Signal processing

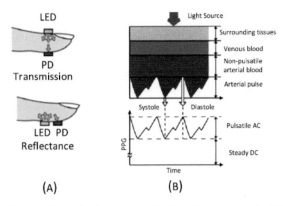

Figure 8.4 (A) Transmission and reflectance-mode photoplethysmography (PPG). (B) Variation in light attenuation by tissue.

technique in time/frequency domain or machine learning models may be used to remove motion artifacts.

Motion artifacts in PPG signal are random and spontaneous action that can be aperiodic or periodic, performing various activities like sitting, walking, or taking the stairs. PPG signals collected in an ambulant environment have motion artifact spectral peaks that are separated from the heart-rate spectral peak, as well as overlapping ones, making them indistinguishable [80]. In some literature, PPG and accelerometers are used together as input to estimate HR. An accelerometer is used to measure and sense motion artifacts of the subject while estimating HR. There are many datasets with only PPG information as primary data for limited period. Attila Reiss *et al.* [84] have introduced a novel, large-scale dataset called *PPG-DaLia*. It includes data recorded for more than 36 hours from 15 subjects by performing various activities close to real-life condition. In this work, the HR estimation has outperformed conventional models using deep learning (convolutional neural network (CNN) technique. The problem with this approach, however, is that it needs heavy computation.

8.5 Studies based on Data Fusion

In this section, we discuss the estimation of clinical information using multi-modal physiological signals. Sometimes, it is exceedingly difficult to estimate based on a single parameter in healthcare, and thus fusion of data with other physiological signal supports estimation in a better way. For instance, fusion of ECG and PPG is used widely for estimation of respiration rate or heartbeat detection in wearable devices [82]. Another popular fusion is EEG and ECG signals for cardiac and brain activity correlation studies. It is observed that the raw EEG signal is mixed with a lot of noise and thus measurements need to be taken with proper care. The one reason for noise in EEG signal is due to motion artifacts of heartbeat, which impacts the signal. This artifact can be removed from EEG signals with proper correlation of ECG signal [82]. ECG, EEG, and electrooculogram (EOG) each have one's own way of representing physiological information and producing satisfactory results even when used exclusively, but they assist in improving performance when fused [83].

Working with fusion data is challenging, as the sampling rate of each sensor varies, and the signal normalization and resampling are part of pre-processing. Most importantly, the measurements must be done simultaneously on subjects without any lag. Data synchronization and pre-processing is imperative before training the model in machine learning.

8.6 Data Pre-processing

In machine learning, data pre-processing is key to designing reliable models. Machine learning algorithms enable the machine to learn from data and improvise without being explicitly programmed. The phrase "garbage in and garbage out" is applicable to machine learning models. Data pre-processing involves transformation of raw data into a meaningful form. Raw data is often incomplete and lack in certain behavior, trend, and/or pattern, which leads to errors in output. Data pre-processing encompassing data cleaning, integration, transformation, reduction, and discretization is a well-proven methodology to resolve such issues [26].

8.7 Conclusion

Thus, machine learning (ML) in healthcare system using electronic health records can be more efficiently mined for different healthcare applications. There is a lot of scope for research in healthcare using EHR. In this chapter, we discussed in detail the various repositories available for study and outlined the systematic approach toward using machine learning techniques on large medical data (EHR) for classification, evaluation, and to improve decision-making in healthcare systems for the betterment of humanity.

In this chapter, we also discussed the use of PPG in biomedical engineering and healthcare. PPG-based heart-rate monitoring became essential in healthcare and fitness applications, as PPG outperforms other methods in estimation and monitoring quality. However, the accuracy of estimation depends on the size of the dataset, with larger ones providing better accuracy. We have also investigated the study on data fusion, i.e., heart-rate detection from fusing various physiological parameters. Finally, it is also established that data pre-processing before fitting the model in machine learning is necessary to ensure that the model development is robust and that it performs as desired.

References

[1] Z. Lipton, D. Kale, C. Elkan, R. W. preprint arXiv:1511.03677, and undefined 2015, "Learning to diagnose with LSTM recurrent neural networks," *arxiv.org*.

[2] U. Desai, R. Martis, C. Nayak, ... K. S.-2015 A. I., and undefined 2015, "Machine intelligent diagnosis of ECG for arrhythmia classification using DWT, ICA and SVM techniques," *ieeexplore.ieee.org*.

[3] S. Kusuma and D. U. J, "Machine Learning and Deep Learning Methods in Heart Disease (HD) Research," vol. 119, no. 18, pp. 1483–1496, 2018.

[4] A. Osareh, B. S.-2010 5th I. S. on, and undefined 2010, "Machine learning techniques to diagnose breast cancer," *ieeexplore.ieee.org*.

[5] P. B. Jensen, L. J. Jensen, and S. Brunak, "Mining electronic health records: Towards better research applications and clinical care," *Nat. Rev. Genet.*, vol. 13, no. 6, pp. 395–405, 2012.

[6] M. R. Boland, N. P. Tatonetti, and G. Hripcsak, "Development and validation of a classification approach for extracting severity automatically from electronic health records," *J. Biomed. Semantics*, vol. 6, no. 1, Apr. 2015.

[7] V. W. Zhong *et al.*, "Use of administrative and electronic health record data for development of automated algorithms for childhood diabetes case ascertainment and type classification: The SEARCH for Diabetes in Youth Study," *Pediatr. Diabetes*, vol. 15, no. 8, pp. 573–584, Dec. 2014.

[8] W. Q. Wei *et al.*, "Evaluating phecodes, clinical classification software, and ICD-9-CM codes for phenome-wide association studies in the electronic health record," *PLoS One*, vol. 12, no. 7, Jul. 2017.

[9] T. Heart, O. Ben-Assuli, and I. Shabtai, "A review of PHR, EMR and EHR integration: A more personalized healthcare and public health policy," *Heal. Policy Technol.*, vol. 6, no. 1, pp. 20–25, 2017.

[10] K. P. Liao *et al.*, "Electronic medical records for discovery research in rheumatoid arthritis," *Arthritis Care Res.*, vol. 62, no. 8, pp. 1120–1127, Aug. 2010.

[11] X. Dong, L. Qian, Y. Guan, L. Huang, Q. Yu, and J. Yang, "A Multiclass Classification Method Based on Deep Learning for Named Entity Recognition in Electronic Medical Records."

[12] M. Saqib, Y. Sha, and M. D. Wang, "Early Prediction of Sepsis in EMR Records Using Traditional ML Techniques and Deep Learning LSTM Networks," *Proc. Annu. Int. Conf. IEEE Eng. Med. Biol. Soc. EMBS*, vol. 2018-July, pp. 4038–4041, 2018.

[13] K. Makantasis, K. Karantzalos, A. Doulamis, and N. Doulamis, *Deep Supervised Learning for Hyperspectral Data Classification Through Convolutional Neural Networks*.

[14] D. Marmanis, M. Datcu, T. Esch, and U. Stilla, "Deep Learning Earth Observation Classification Using ImageNet Pretrained Networks," *IEEE Geosci. Remote Sens. Lett.*, vol. 13, no. 1, 2016.

[15] Z. Liang, G. Zhang, J. X. Huang, and Q. V. Hu, "Deep learning for healthcare decision making with EMRs," *Proc. - 2014 IEEE Int. Conf. Bioinforma. Biomed. IEEE BIBM 2014*, no. Cm, pp. 556–559, 2014.

[16] H. Act, "Health Information Technology for Economic and Clinical Health," 2010.

[17] B. Shickel, P. J. Tighe, A. Bihorac, and P. Rashidi, "Deep EHR: A Survey of Recent Advances in Deep Learning Techniques for Electronic Health Record (EHR) Analysis," *IEEE J. Biomed. Heal. Informatics*, vol. 22, no. 5, pp. 1589–1604, 2018.

[18] A. Radford, L. Metz, and S. Chintala, "Unsupervised Representation Learning with Deep Convolutional Generative Adversarial Networks," Nov. 2015.

[19] Y. Bengio, A. Courville, P. V.-I. transactions on pattern, and undefined 2013, "Representation learning: A review and new perspectives," *ieeexplore.ieee.org*.

[20] B. L. P. Cheung and D. Dahl, "Deep learning from electronic medical records using attention-based cross-modal convolutional neural networks," *2018 IEEE EMBS Int. Conf. Biomed. Heal. Informatics, BHI 2018*, vol. 2018-Janua, no. March, pp. 222–225, 2018.

[21] S. Lim, C. S. Tucker, and S. Kumara, "An unsupervised machine learning model for discovering latent infectious diseases using social media data," *J. Biomed. Inform.*, vol. 66, pp. 82–94, 2017.

[22] "diagnosing abnormalities using EMR - Google Scholar." [Online]. Available: https://scholar.google.com/scholar?hl=en&as_sdt=0%2C5&q=diagnosing+abnormalities+using+EMR&btnG=. [Accessed: 10-Dec-2019].

[23] H. Singh *et al.*, "HEALTH CARE REFORM Timely Follow-up of Abnormal Diagnostic Imaging Test Results in an Outpatient Setting Are Electronic Medical Records Achieving Their Potential?"

[24] H. Singh, A. D. Naik, R. Rao, and L. A. Petersen, "Reducing diagnostic errors through effective communication: Harnessing the power of information technology," *Journal of General Internal Medicine*, vol. 23, no. 4. pp. 489–494, Apr-2008.

[25] K. Häyrinen, K. Saranto, P. N.-I. journal of medical, and undefined 2008, "Definition, structure, content, use and impacts of electronic health records: a review of the research literature," *Elsevier*.

[26] S. B. Kotsiantis, "Supervised Machine Learning: A Review of Classification Techniques," 2007.

[27] D. D. Wang, S. H. X. Ng, S. N. B. Abdul, S. Ramachandran, S. Sridharan, and X. Q. Tan, "Imputation of Missing Diagnosis of Diabetes in an Administrative EMR System," *BMEiCON 2018 - 11th Biomed. Eng. Int. Conf.*, pp. 1–5, 2019.

[28] H. Huang *et al.*, "Discovering Medication Patterns for High-Complexity Drug-Using Diseases Through Electronic Medical Records," *IEEE Access*, vol. 7, pp. 125280–125299, 2019.

[29] M. Li, Y. Zhang, M. Huang, J. Chen, and W. Feng, "Named Entity Recognition in Chinese Electronic Medical Record Using Attention Mechanism," *2019 Int. Conf. Internet Things IEEE Green Comput. Commun. IEEE Cyber, Phys. Soc. Comput. IEEE Smart Data*, pp. 649–654, 2019.

[30] S. Nemati, A. Holder, F. Razmi, M. D. Stanley, G. D. Clifford, and T. G. Buchman, "An Interpretable Machine Learning Model for Accurate Prediction of Sepsis in the ICU," *Crit. Care Med.*, vol. 46, no. 4, pp. 547–553, 2018.

[31] S. Zhang, S. M. H. Bamakan, Q. Qu, and S. Li, "Learning for Personalized Medicine: A Comprehensive Review from a Deep Learning Perspective," *IEEE Rev. Biomed. Eng.*, vol. 12, pp. 194–208, 2018.

[32] D. Chen, G. Qian, and Q. Pan, "Breast Cancer Classification with Electronic Medical Records Using Hierarchical Attention Bidirectional Networks," *2018 IEEE Int. Conf. Bioinforma. Biomed.*, pp. 983–988, 2018.

[33] K. H. Miao and J. H. Miao, "Coronary heart disease diagnosis using deep neural networks," *Int. J. Adv. Comput. Sci. Appl.*, vol. 9, no. 10, pp. 1–8, 2018.

[34] Z. Huang, T. M. Chan, and W. Dong, "MACE prediction of acute coronary syndrome via boosted resampling classification using electronic medical records," *J. Biomed. Inform.*, vol. 66, pp. 161–170, 2017.

[35] A. Dey, J. Singh, and N. Singh, "Analysis of Supervised Machine Learning Algorithms for Heart Disease Prediction with Reduced

Number of Attributes using Principal Component Analysis," *Int. J. Comput. Appl.*, vol. 140, no. 2, pp. 27–31, 2016.

[36] R. J. Byrd, S. R. Steinhubl, J. Sun, S. Ebadollahi, and W. F. Stewart, "Automatic identification of heart failure diagnostic criteria, using text analysis of clinical notes from electronic health records," *Int. J. Med. Inform.*, vol. 83, no. 12, pp. 983–992, Dec. 2014.

[37] M. W. A. Caan and M. Welling, "Recurrent inference machines for reconstructing heterogeneous MRI data R," vol. 53, pp. 64–78, 2019.

[38] C. Zhao, J. Jiang, Y. Guan, X. Guo, and B. He, "Artificial Intelligence in Medicine EMR-based medical knowledge representation and inference via Markov random fields and distributed representation learning," *Artif. Intell. Med.*, vol. 87, pp. 49–59, 2018.

[39] C. Zhao, J. Jiang, Z. Xu, and Y. Guan, "Computer Methods and Programs in Biomedicine A study of EMR-based medical knowledge network and its applications," *Comput. Methods Programs Biomed.*, vol. 143, pp. 13–23, 2017.

[40] J. Jiang, X. Li, C. Zhao, Y. Guan, and Q. Yu, "Knowledge-Based Systems Learning and inference in knowledge-based probabilistic model for medical diagnosis," vol. 138, pp. 58–68, 2017.

[41] I. Kononenko, "Machine learning for medical diagnosis: History, state of the art and perspective," *Artif. Intell. Med.*, vol. 23, no. 1, pp. 89–109, 2001.

[42] S. Redmond, Q. Lee, ... Y. X.-... C. of the I., and undefined 2012, "Applications of supervised learning to biological signals: ECG signal quality and systemic vascular resistance," *ieeexplore.ieee.org*.

[43] D. Hübner, T. Verhoeven, K. R. Müller, P. J. Kindermans, and M. Tangermann, "Unsupervised Learning for Brain-Computer Interfaces Based on Event-Related Potentials: Review and Online Comparison [Research Frontier]," *IEEE Comput. Intell. Mag.*, vol. 13, no. 2, pp. 66–77, 2018.

[44] "Machine Learning - 2nd Edition." [Online]. Available: https://www.elsevier.com/books/machine-learning/theodoridis/978-0-12-818803-3. [Accessed: 20-Nov-2019].

[45] V. Narayanan, I. Arora, and A. Bhatia, "Fast and Accurate Sentiment Classification Using an Enhanced Naive Bayes Model," 2013, pp. 194–201.

[46] T. Calders and S. Verwer, "Three naive Bayes approaches for discrimination-free classification," *Data Min. Knowl. Discov.*, vol. 21, no. 2, pp. 277–292, Sep. 2010.

[47] D. Montgomery, E. Peck, and G. Vining, *Introduction to linear regression analysis*. 2012.

[48] G. Seber and A. Lee, *Linear regression analysis*. 2012.

[49] K. W. Johnson *et al.*, "Artificial Intelligence in Cardiology," *J. Am. Coll. Cardiol.*, vol. 71, no. 23, pp. 2668–2679, 2018.

[50] F. Amato, A. López, E. Peña-Méndez, and P. Vaňhara, "Artificial neural networks in medical diagnosis," 2013.

[51] G. Daniel, *Principles of artificial neural networks*. 2013.

[52] J. Wang, X. S.-2011 I. 3rd I. C. on, and undefined 2011, "An improved K-Means clustering algorithm," *ieeexplore.ieee.org*.

[53] V. Dehariya, … S. S.-2010 I., and undefined 2010, "Clustering of image data set using k-means and fuzzy k-means algorithms," *ieeexplore.ieee.org*.

[54] C. Morais, K. L.-J. of the B. C. Society, and undefined 2018, "Principal component analysis with linear and quadratic discriminant analysis for identification of cancer samples based on mass spectrometry," *SciELO Bras.*

[55] C. Anagnostopoulos, … D. T.-S. A., and undefined 2012, "Online linear and quadratic discriminant analysis with adaptive forgetting for streaming classification," *Wiley Online Libr.*

[56] S. Bose, A. Pal, R. SahaRay, J. N.-P. Recognition, and undefined 2015, "Generalized quadratic discriminant analysis," *Elsevier*.

[57] Y. Ma and G. Guo, *Support vector machines applications*. 2014.

[58] D. Meyer, F. W.-T. I. to libsvm in package e1071, and undefined 2015, "Support vector machines," *cse.yzu.edu.tw*.

[59] N. Suguna, K. T.-I. J. of Computer, and undefined 2010, "An improved k-nearest neighbor classification using genetic algorithm," *research-gate.net*.

[60] I. Saini, D. Singh, A. K.-J. of advanced research, and undefined 2013, "QRS detection using K-Nearest Neighbor algorithm (KNN) and evaluation on standard ECG databases," *Elsevier*.

[61] N. Bhatia and Vandana, "Survey of Nearest Neighbor Techniques," Jul. 2010.

[62] R. Barros, M. B.-… on Systems, undefined Man, and undefined 2011, "A survey of evolutionary algorithms for decision-tree induction," *ieeexplore.ieee.org*.

[63] B. Hssina, A. Merbouha, … H. E.-I. J. of, and undefined 2014, "A comparative study of decision tree ID3 and C4. 5," *Citeseer*.

[64] H. Assodiky, I. Syarif, and T. Badriyah, "Deep learning algorithm for arrhythmia detection," *Proc. - Int. Electron. Symp. Knowl. Creat. Intell. Comput. IES-KCIC 2017*, vol. 2017-Janua, pp. 26–32, 2017.

[65] L. Deng and D. Yu, "Deep Learning: Methods and Applications Foundations and Trends R in Signal Processing," *Signal Processing*, vol. 7, no. 2013, pp. 197–387, 2013.

[66] A. Voulodimos, N. Doulamis, … A. D.-C., and undefined 2018, "Deep learning for computer vision: A brief review," *hindawi.com.*

[67] M. Buyukyilmaz, A. C.-2016 I. Conference, and undefined 2016, "Voice gender recognition using deep learning," *atlantis-press.com.*

[68] T. Young, D. Hazarika, … S. P. C., and undefined 2018, "Recent trends in deep learning based natural language processing," *ieeexplore.ieee.org.*

[69] A. Krizhevsky, I. Sutskever, and G. E. Hinton, "ImageNet Classification with Deep Convolutional Neural Networks." pp. 1097–1105, 2012.

[70] Z. Lin, Y. Gu, Y. Chen, X. Zhao, and G. Wang, "Deep Learning-Based Classification of Hyperspectral Data," *Artic. IEEE J. Sel. Top. Appl. Earth Obs. Remote Sens.*, 2014.

[71] L. Nie, M. Wang, L. Zhang, … S. Y.-I. T. on, and undefined 2015, "Disease inference from health-related questions via sparse deep learning," *ieeexplore.ieee.org.*

[72] A. Esteva *et al.*, "Dermatologist-level classification of skin cancer with deep neural networks," *Nature*, vol. 542, no. 7639, pp. 115–118, 2017.

[73] E. Putin, P. Mamoshina, A. Aliper, … M. K.-A. (Albany, and undefined 2016, "Deep biomarkers of human aging: application of deep neural networks to biomarker development," *ncbi.nlm.nih.gov.*

[74] N. J. Ravindran and P. Gopalakrishnan, "Big data Technology," *2018 Second Int. Conf. Green Comput. Internet Things*, no. Ml, pp. 326–331, 2018.

[75] M. R. Ahmed, Y. Zhang, Z. Feng, B. Lo, O. T. Inan, and H. Liao, "Neuroimaging and Machine Learning for Dementia Diagnosis: Recent Advancements and Future Prospects," *IEEE Rev. Biomed. Eng.*, vol. 12, pp. 19–33, 2019.

[76] M. Ahmed, Y. Zhang, Z. Feng, … B. L.-I. reviews in, and undefined 2018, "Neuroimaging and Machine Learning for Dementia Diagnosis: Recent Advancements and Future Prospects," *ieeexplore.ieee.org.*

[77] U. Desai, R. J. Martis, C. G. Nayak, K. Sarika, and G. Seshikala, "Machine intelligent diagnosis of ECG for arrhythmia classification using DWT, ICA and SVM techniques," *12th IEEE Int. Conf. Electron. Energy, Environ. Commun. Comput. Control (E3-C3), INDICON 2015*, pp. 1–4, 2016.

[78] H. Assodiky, I. Syarif, and T. Badriyah, "Deep learning algorithm for arrhythmia detection," in *2017 International Electronics Symposium on Knowledge Creation and Intelligent Computing (IES-KCIC)*, 2017, pp. 26–32.

[79] R. Poplin, A. Varadarajan, K. Blumer, … Y. L.-N. B., and undefined 2018, "Prediction of cardiovascular risk factors from retinal fundus photographs via deep learning," *nature.com.*

[80] https://www.researchgate.net/publication/327287312_Bio-Signal_ System_Design_for_Real-Time_Ambulatory_Patient_Monitoring_ and_Abnormalities_Detection_System_Proceedings_of_the_6th_ ICIECE_2017

[81] Z. He, X. Chen, Z. Fang, T. Sheng and S. Xia, "Fusion estimation of respiration rate from ECG and PPG signal based on Android platform and wearable watch," 2nd IET International Conference on Biomedical Image and Signal Processing (ICBISP 2017), 2017, pp. 1-6, doi: 10.1049/cp.2017.0096.

[82] R. K. C. Billones et al., "Cardiac and Brain Activity Correlation Analysis Using Electrocardiogram and Electroencephalogram Signals," 2018 IEEE 10th International Conference on Humanoid, Nanotechnology, Information Technology,Communication and Control, Environment and Management (HNICEM), 2018, pp. 1-6, doi: 10.1109/ HNICEM.2018.8666392.

[83] Tejedor J, García CA, Márquez DG, Raya R, Otero A. Multiple Physiological Signals Fusion Techniques for Improving Heartbeat Detection: A Review. Sensors (Basel). 2019;19(21):4708. Published 2019 Oct 29. doi:10.3390/s19214708

[84] "Deep PPG: Large-Scale Heart-rate Estimation with Convolutional Neural Networks". Sensors 2019, 19(14), 3079; https://doi.org/10.3390/ s19143079

[85] D. Biswas, N. Simões-Capela, C. Van Hoof and N. Van Helleputte, "Heart Rate Estimation From Wrist-Worn Photoplethysmography: A Review," in IEEE Sensors Journal, vol. 19, no. 16, pp. 6560-6570, 15 Aug.15, 2019, doi: 10.1109/JSEN.2019.2914166.

[86] F. Fritz, B. Tilahun, and M. Dugas, "Success criteria for electronic medical record implementations in low-resource settings: A systematic review," Journal of the American Medical Informatics Association, vol. 22, no. 2, pp. 479–488, Mar. 2015, doi: 10.1093/JAMIA/OCU038.

[87] T. Heart, O. Ben-Assuli, and I. Shabtai, "A review of PHR, EMR and EHR integration: A more personalized healthcare and public health policy," Health Policy and Technology, vol. 6, no. 1, pp. 20–25, Mar. 2017, doi: 10.1016/J.HLPT.2016.08.002.

[88] R. J Johnson III, "A Comprehensive Review of an Electronic Health Record System Soon to Assume Market Ascendancy: EPIC®," Journal of Healthcare Communications, vol. 01, no. 04, 2016, doi: 10.4172/2472-1654.100036.

[89] R. S. Evans, "Electronic Health Records: Then, Now, and in the Future," Yearbook of medical informatics, pp. S48–S61, May 2016, doi: 10.15265/IYS-2016-S006.

[90] S. Biruk, T. Yilma, M. Andualem, and B. Tilahun, "Health Professionals' readiness to implement electronic medical record system at three hospitals in Ethiopia: A cross sectional study," BMC Medical Informatics and Decision Making, vol. 14, no. 1, Dec. 2014, doi: 10.1186/S12911-014-0115-5.

[91] S. H. Afrizal, A. N. Hidayanto, P. W. Handayani, M. Budiharsana, and T. Eryando, "Narrative Review for Exploring Barriers to Readiness of Electronic Health Record Implementation in Primary Health Care," Healthcare Informatics Research, vol. 25, no. 3, pp. 141–152, Jul. 2019, doi: 10.4258/HIR.2019.25.3.141.

[92] S. Y. Shin and C. K. Chung, "A future of medical information system: Establishment of hospital-oriented Personal Health Record," Journal of the Korean Medical Association, vol. 52, no. 11, pp. 1115–1121, Nov. 2009, doi: 10.5124/JKMA.2009.52.11.1115.

[93] Y. Park and H.-J. Yoon, "Understanding Personal Health Record and Facilitating its Market," Healthcare Informatics Research, vol. 26, no. 3, pp. 248–250, Jul. 2020, doi: 10.4258/HIR.2020.26.3.248.

9

Machine Learning-based Integrated Approach for Cancer Microarray Data Analysis

Amrutanshu Panigrahi[1], Manoranjan Dash[2], Bibhuprasad Sahu[3], Abhilash Pati[4], and Sachi Nandan Mohanty[5]

[1]Department of CSE, Siksha 'O' Anusandhan (Deemed to be University), India.
[2]Department of Management Sciences, Siksha 'O' Anusandhan (Deemed to be University), India.
[3]Department of Computer Science and Engineering, Gandhi Institute for Technology, India.
[4]Department of CSE, Siksha 'O' Anusandhan (Deemed to be University), India.
[5]VIT-AP University, Amaravati, AP, India.
Email: amrutansup89@gmail.com; manoranjandash@soa.ac.in; prasadnikhil176@gmail.com; er.abhilash.pati@gmail.com; sachinandan@ieee.org

Abstract

Cancer is increasingly becoming the leading cause of death around the world. Machine learning is critical in the deployment of an automated model that aids in better diagnosis. The traditional cancer diagnosis relies solely on biopsy data, which overlooks essential elements of disease such as the rate of proliferation and tissue behavior. As a result, genetic data may play an essential role in identifying cancer subtypes. Microarray data contains a patient's genetic information in a large number of dimensions, such as genes, with limited sample size, such as patient details. If the microarray is directly taken without reducing the dimension as the input to any ML model for classification, then small sample size is the resulting issue. So, the microarray data has to be normalized by using either the dimensionality reduction technique or the feature selection technique. The main objective of this research work is to analyze the impact of the microarray dataset in cancer classification. The next focus is to study the various kinds of machine learning algorithms that can be used for cancer microarray data along with various validation methods available in calculating the accuracy of the algorithm. In the current research, the main focus is on building an integrated approach based on feature selection algorithm, optimization algorithm, and machine learning classification algorithm for efficient cancer classification. For this research work, we will utilize RFE for dimensionality reduction, cuckoo search (CS) for optimization, and SVM for classification for multiple benchmark datasets. The performance of the proposed model will be evaluated based on some tuned parameters such as accuracy, sensitivity, specificity, and $F1$-score. Finally, the result will be compared with state-of-the-art machine learning algorithms.

9.1 Introduction

In the current era, machine learning is playing a vital role in the healthcare system. The conventional system for detecting the disease is to verify the biopsy dataset, which misses the various important parameters such as the genetic information. An accurate model can help in increasing the recovery rate. As per the WHO, the number of deaths is increasing and machine learning gives an appropriate solution. As per the NCI, around 12% of lung cancer rises every year, out of which 10% of the patient die. If the case of breast cancer is being considered, then around 11% of new patients come across the world, out of which 9% die. For handling cancer, it is required that the correct and meaningful dataset needs to be generated [1]. There are three different kinds of datasets available, such as the clinical data, omics data, and sensor data. The clinical data is also known as the biopsy data, which will contain the test results of the patients. The conventional system for detecting the disease uses this kind of dataset. The main issue present behind this kind of dataset is that it misses various important information such as genetic information of the patient, which can help in predicting the disease more accurately [2]. To overcome this limitation, we have an omics dataset, which is also known as the microarray data. It will contain the genetic information of the patient, which can help the diagnosis system to detect the disease more accurately. The microarray data also has the issue of high dimensionality as the dataset contains the huge number of genetic information as compared to the number of samples [3]. So if any kind of machine learning algorithm is applied to the omics dataset, then the model will face the problem known as the small sample size. To overcome this problem, the dimension needs to be reduced by using any kind of reduction technique. The dimensionality reduction technique can be classified into two groups such as feature selection, which will remove the irrelevant gene expression data, and feature extraction, which will add some features to a few features from a large number of original features through their linear and nonlinear combination [4]. The sensor data will contain the real-time data that is being collected from the wearable device connected with the patient. This dataset will generally contain the pulse rate, blood pressure, etc. But this data cannot be taken to build a model for disease classification and detection. In recent years, the production and analysis of microarray data have increased their importance due to achieving better accuracy in the diagnosis and classification of cancer disease [1].

Disease prediction systems are extremely important since they are responsible for determining whether or not a person has a medical condition. It entails a variety of features, qualities, and complex and real-world components [1, 2]. In recent years, there has been a growing demand for data-driven and reliable prediction models to improve the precision of future event identification [3]. Cancer screening recommendations and guidance are available from several medical associations and patient counseling programs. But the microarray technology produces huge numbers of genes for fewer samples at a time. From these genes, some are not relevant in the classification of the disease. As the microarray data is having a high number of features, noise, and computation complexity for a small sample size, it is not easy to classify the microarray data. There are various feature selection and classification techniques proposed to deal with these limitations. But none of the feature selection techniques or classification algorithms work better for the entire microarray datasets [2]. Hence, to explore new hybrid approaches, further studies should be conducted to get efficient results [3]. To enhance the microarray dataset classification, different feature selection and classification algorithms are designed. Most of the analyses are done by using breast [4], colon [5, 6], ovarian [7], and leukemia datasets. Microarray data production and processing have become more important in recent years as a means of improving cancer disease detection and classification accuracy [1]. However, microarray technology generates a large number of genes for a small number of samples at a time. Some of these genes have no bearing on the disease classification. Microarray data is difficult to categorize because of the large number of characteristics, noise, and computational complexity for such a tiny sample size. To address these restrictions, different feature selection and classification strategies have been developed.

A huge amount of gene expression data are produced in a single trial on microarray data. MT presents an enormous chance in determining disease association with genes. But, the gene expression data are distinguished with high dimensionalities that are unimportant in diagnosing diseases. The high dimensionality gene expression data are noisy, redundant, and irrelevant. The reason for deprived diagnosis as well as classification is due to the ratio of a huge number of genes (features) and less number of patients (samples) with the presence of redundant data gene expression. It can shortly narrate the limitations of microarray data, which are directly affecting the classification accuracy; inability to process each data every time and its subset data processing may lead to the reasons behind over-fitting, information loss, etc. The most common benefits of feature selection are that it improves accuracy and reduces over-fitting and the time required for training. To address the factors related to dimensionality, there have been a lot of research works done in identifying features that have high impacts. As the irrelevant features should be reduced and needful information should be extracted from the microarrays, the feature selection methods do the same. Also, the feature selection methods list the attributes that are the main causes of the diseases. The primary objective of feature selection is to pick a subset of features from the inputted data. From various research works, it is observed that the redundant features have no such impacts on classification problems. Hence, these should be reduced to enhance the classification problems. In addition, the noisy data, features that do not have any correlation with classes totally, should be diminished to improve classification accuracy [3].

The goals of feature selection can be listed as follows:

i)   diminishes the noisy features to improve the classification accuracy;

ii) removes the redundant data to enhance the classification problem;

iii) reduces the irrelevant data to process faster;

The structure of the chapter is as follows. Section 9.2 holds the literature survey done during the research work. The background study is mentioned in Section 9.3. Section 9.4 shows the proposed work and Section 9.5 holds the empirical analysis of the proposed work. Finally, the conclusion is given in Section 9.6.

9.2 Related Work

In [13], the author has used SVM as the classifier to the microarray data and found the classification accuracy as 97.802%. Sharma *et al.* [14] have used gradient LDA for the classification and LDA technique as the dimensionality reduction and obtained 100% accuracy with the microarray dataset. In [15], SVM has been utilized as the classifier technique with a genetic algorithm and the proposed model achieves a 98.28% accuracy level. In [16], PNN in addition to ICA and discrete wavelet-based feature selection algorithm has been used to cancer microarray data and the proposed model gave a 96.02% of accuracy level. The authors in [17] proposed a hybrid model based on SVM and PSO with PCA dimensionality reduction technique and obtained a 97.02% of accuracy level. In [18], the author used SVM for the classification purpose and ICA as the dimensionality reduction method to a microarray dataset and obtained 94.4% accuracy. In [19], the author had proposed an ensemble method based on SPLSDA classification, binary black hole optimization, and binary PSO and showed that a 100% accuracy level has been achieved for the cancer microarray dataset.

In [20], mRMR-ABC-SVM model has been proposed and the result shows that the accuracy level is 100% by using the proposed model. Jain *et al.* [21] have used a hybrid approach Relief-PCA for feature selection purposes and upon the featured dataset, the model achieves a 98.5% of accuracy level. Gupta *et al.* [22] had implemented an ensemble method with RNN and CNN as the classifier with mRMR, PCA, and LDA dimensionality reduction technique and found that the accuracy during the performance analysis reaches up to 96.58%. In [23], the author had taken NB as the classifier with correlation-based feature selection (CFS) and traditional BPSO as the dimensionality reduction technique, which has been implemented in 11 benchmark datasets, and showed that the proposed CFS-BPSO-NB achieves an average accuracy of 100%. Mabarti *et al.* [24] have implemented a hybrid model based on the DT and GA as the classifier and optimization technique. For dimensionality reduction purposes, the author had used mRMR and got 78% accuracy during the performance analysis. In [25], SVM-CS hybrid model has been proposed based on the mRMR dimensionality reduction technique with an accuracy of 67%.

Astuti *et al.* [26] have used the NN-ABC hybrid model with RF and PCA dimensionality reduction technique with the cancer dataset and got the accuracy as 96.8%. In [27], the author had proposed one more hybrid model based on SVM and PCA with a 96.07% accuracy level. The author in [28] had proposed a DT-BPSO hybrid model for breast cancer classification with 99% accuracy. Kavita *et al.* [29] had proposed the SVM-RFE-PCA ensemble method for lung cancer classification and considered the microarray dataset for analysis purposes and found the accuracy as 97%. The author in [30] has proposed the SVM-GA-PCA-PLS ensemble method for obtaining an accuracy of 88.49%. Table 9.1 describes the overall finding of the literature survey done during the research work along with the corresponding performance measure.

9.3 Background Study

In this section, the main objective is to study the concept of dimensionality reduction techniques along with the pros and cons. Also, the microarray dataset along with its research challenges and the available classification technique has been focused on.

9.3.1 Machine learning

Machine learning gives an automatic approach for solving real-time tasks based on classification and prediction approaches. Both approaches of machine learning require two phases as the training and testing phase. In the training phase, some attributes of the dataset will be provided to the machine learning algorithm for understanding purposes, and depending upon the training phase, the algorithm gives the solution. Higher the training set, better will be the result. Instead of the similarity, there is a thin line difference present between these two approaches. The classification method tries to classify the samples based on the attributes present in the dataset, whereas the prediction aims to provide some future probable decisions. In this survey work, the main focus is on the classification approach.

9.3.1.1 Types of machine learning algorithms

There are three major categories of machine learning algorithms:

- Supervised learning: During training, the machine is given labeled data for both inputs and expected output, and the supervised learning algorithm creates a mapping function that can identify the predicted output for a given input. The training process is repeated until the algorithm achieves the desired degree of precision. Because one of the specified aims of supervised learning is to teach a computer a classification system, it is frequently employed to tackle classification difficulties. For example, the machine may be trained to distinguish between spam and legitimate e-mails, as Google now does for Gmail spam filtering.

- Unsupervised learning: The machine is given an unlabeled and unclassified input dataset, and the unsupervised learning method develops a function to discover hidden structures in the dataset based on the

Table 9.1 Related works.

Ref.	Classifier	Optimization technique	Dataset	Dimensionality reduction technique	Accuracy
[13]	SVM	–	Microarray dataset	–	97.802
[14]	GLDA	–	Microarray data	LDA	100
[15]	SSVM + SVM	GA	Microarray data	–	98.28
[16]	PNN	–	Microarray data	ICA + DWT	96.02
[17]	SVM	PSO	Microarray data	PCA	97.02
[18]	SVM	–	Microarray data	ICA	94.4
[19]	SPLSDA	BBH	Microarray data	BPSO	100
[20]	SVM	ABC	Microarray data	MRMR	100
[21]	SVM	–	Microarray data	Relief-PCA	98.5
[22]	RNN + CNN	–	Microarray data	mRMR + PCA + LDA	96.58
[23]	NB	–	Microarray data	CFS + iBPSO	100
[24]	C4.5 DT	GA	Microarray data	mRMR	78
[25]	SVM	CS	Microarray data	mRMR	67
[26]	SVM	–	Microarray data	PCA	100
[27]	SVM + LMBP	–	Microarray data	PCA	94.08/96.07
[28]	C4.5 DT	–	Microarray data	BPSO	99
[29]	SVM-RFE	–	Microarray data	PCA	96.44
[30]	SVM	GA	Microarray data	PCA+PLS	88.49

patterns, similarities, and differences that exist among the data without any prior training. There is no assessment of the machine's accuracy in identifying the structure. Clustering and association problems could be one of the main foci of unsupervised learning methods. The *k*-means technique for clustering and the Apriori algorithm for association problems are two examples of widely used unsupervised learning algorithms.

- Reinforcement learning: The machine is placed in a situation where it must make judgments based on trial and error and learn from its own actions and previous experiences. The computer receives a reward input from the environment for each right decision, which works as a reinforcement signal, and the knowledge about the rewarded state-action combination is recorded. When confronted with a similar situation again, the machine repeats the rewarded behavior. Reinforcement learning algorithms are used in fields like self-driving cars, where strategic decision-making is critical to success. Here are a few of the most widely utilized ones.

9.3.1.2 Machine learning algorithms

One of the paths to knowledge begins with perceptions or information, such as precedents, coordinated involvement, or direction, with the purpose of finding examples in data and making better decisions later on based on the models we provide. The main goal is to allow the PCs to grasp organically without the need for human intervention or assistance and to change their behaviors in the same way. Basically, in machine learning, there are seven well-known classification algorithms present, which are as follows:

- Support vector machine: It is a supervised machine learning approach that aims to locate a hyper-plane in a multidimensional space. The SVM's goal is to find the best hyper-plane between the two classes and separate them. This ideal hyper-plane divides the two classes while also increasing the margin between them. The gap between the hyper-plane and the SVs is known as the margin. It is very popular due to its primary advantage, i.e., it can be a very effective one even in the high-dimensional space, but the main lacuna present behind this approach is that it does not provide probabilistic estimations. The high accuracy can be achieved by tuning the hyperparameters such as gamma, coat, and kernel-level, but in reality, it becomes very hard to define the exact hyperparameters, which directly enhances the computational cost and overhead.

- Decision tree: Given a set of attributes, the decision tree tries to create a rule based on the classification that can be done. In a decision tree generally, the process is being started with a root and gradually the data attributes are divided based on the information gain score. The DT will contain three types of nodes known as the root node, test node, and leave node known as the decision node. A DT is being constructed by using some classifier such as ID3, CART, CHAID, etc., which generally defines a tree automatically. But the main objective of the classification algorithm is to construct an optimized decision tree for the given dataset. It makes the utilization of different algorithms to make a decision as to whether one node can be divided into two or more sub-nodes. The main advantage of DT is that it requires very less data pre-processing before implementation and is also able to handle both the numerical and categorical data. But over-fitting is the main disadvantage of this classification technique.

- Naïve Bayes: The naïve Bayes (NB) classification algorithm works on the basis of Bayes conditional probability theorem. In this, probability means the degree of belief. The conditional probability is being used to classify the data. The most important part of this algorithm is that it works with the assumption that all of the attributes are independent of each other. There are three different kinds of NB-based algorithms present as the Gaussian NB, multinominal NB, and Bernoulli NB. The main advantage present behind this classification is that it requires a very small amount of training data for estimating the conditional parameters, but the estimation time depends upon the dataset size. If the dimension of the dataset is very high, then the NB will be acting as a bad estimator as the estimation time and cost increase with respect to the dataset.

- Neural network: The learning interaction that occurs in the human brain motivates neural networks. They include a forgery capability that allows the computer to learn and calibrate itself by examining new data. Every parameter, also known as a neuron, is a function that produces an output after receiving data from one or more sources. These outputs are subsequently passed on to the next layer of neurons, which use them as inputs to their own processes and generate more output.

- Logistic regression: It is the regression model that can be used to predict the probability of a given data. Its working depends on the well-defined model known as the logistic function also known as the sigmoid function. In this model, the probabilities define the possible outcomes of a single trial tuned by the sigmoid function. The main advantage of this classification model is that it can be easily implemented on independent variables but finding out the independent variable in a high-dimensional data.

- *k*-NN: Neighbors-based order is a sort of languid learning as it does not endeavor to build an overall inside model; however, it essentially stores cases of the preparation information. Grouping is processed from a straightforward larger part vote of the *k* closest neighbors of each point. The main advantages of this classification algorithm are that it is very easy to implement and also it is robust to the noise present in the training data. The disadvantages of this model include the determination of the *k* value as the improper *k* value can intend low performance.

- Random forest: Random forest is one of the most prominent ensemble techniques, which makes the utilization of multiple decision trees. The diversity is obtained by randomizing the selection criteria of the node split. In this, any random feature is chosen instead of choosing the best feature for splitting purposes. It is a bagging approach where deep trees are combined together to form a low variance output. When it receives an input *x*, where *x* is a vector made up of various features, then the RF performs the *n* time decision tree execution and then makes an average of the result to find out the actual prediction.

9.3.2 Microarray data

Microarray data are the large source of the genetic information upon which a better machine learning model can be built up. The amount of data necessary to perform a credible analysis grows exponentially as the complexity of the data increases in machine learning. When considering challenges in dynamic optimization, Bellman refers to this phenomenon as the "curse of dimensionality" [31, 32]. Searching for a projection of the data onto a smaller number of variables (or features) that maintain as much information as feasible is a prominent approach to this problem of high-dimensional datasets. With the advent of microarray technology, the researchers are able to create a change in different disease diagnoses and prognoses in contrast to the non-genetic dataset or biopsy dataset [34]. But the main issue while dealing with the microarray dataset is the high dimension of the dataset. The microarray dataset contains a huge amount of genetic information that can play an important role in dealing with different disease prognoses.

Different types of microarray data are present, which are described in the following [35].

- DNA microarray: It is also known as the gene chip data or biochip data that measures the behavior of the DNA; there are four kinds of DNA microarray datasets, such as the cDNA, oligo microarray, BAC microarray dataset, and SNP microarray dataset [36].

- MM chips: MM chip data enable the researcher to analyze the cross-platform and the laboratory data. It also helps in finding the relationship between DNA and protein.

- Protein microarray: It is a platform for analyzing multiple proteins in an effective parallel way. It can further be divided into three different subcategories such as the protein, functional protein, and reverse-phase protein microarray data.

- Peptide microarray: This microarray data is used to determine the interaction between the proteins, which can enable the researcher to detect the antibody by screening the appropriate proteomes.

- Phenotype microarray: This microarray analysis is being used to develop corresponding drugs for different kinds of diseases. The main objective of analyzing this kind of microarray is to recognize the characteristic of the particular disease and to suggest the appropriate antidote that can be applied for diagnosing that corresponding disease [37].

9.4 Proposed Work

In the current research, the main focus is on building an integrated approach based on feature selection algorithm, optimization algorithm, and machine learning classification algorithm for efficient cancer classification. For this research work, we will utilize RFE for dimensionality reduction, cuckoo search (CS) for optimization, and SVM for classification of multiple benchmark datasets. Figure 9.1 shows the proposed work.

9.4.1 RFE

The RFE selection method [11] is essentially a recursive procedure that ranks features based on some metric of relevance. At each iteration, the importance of each feature is assessed, and the one that is not very important is removed. Another option, which was not used in this case, is to eliminate a group of characteristics at a time to speed up the process. Because the relative value of each characteristic can change significantly when examined over a different selection of features during the stepwise elimination procedure for various metrics, recursion is required. The final ranking is based on the (inverse) order in which features are eliminated. Only the first n characteristics from this rating are used in the feature selection procedure.

Let T be the training set with the number of features as $F = \{f1, f2,...,$ $fn\}$. The ranking function is denoted by Rank () with T and F as the input. The rank is denoted by R. The pseudo-code for RFE is given as follows.

```
for i = {1 → n}
     R = Rank (T, F)
     f' = The last ranked feature among F
     R (n − i + 1) = f'
     F = F − f'.
```

9.4.2 Cuckoo search

The cuckoo search method is a nature-inspired algorithm based on cuckoo bird reproduction [14]. It is crucial to correlate prospective solutions with cuckoo eggs while working with CS algorithms. Cuckoos usually lay their fertilized eggs in the nests of other cuckoos in the hopes that their offspring would be nurtured by proxy parents. When the cuckoos find that the eggs in their nests do not belong to them, the foreign eggs are either tossed out of the nests or the nests are abandoned entirely. The pseudo-code for the CS algorithm is as follows:

Start

    Objective function $F(n)$, $n = (n_1,..., n_x)$

    Initial population of p host nests n_i $(i = 1, 2,..., p)$;

    Get a cuckoo and produce a new solution through Lévy flights;

    Calculate fitness F_i

    Choose a nest among n (j) randomly;

    if $(F_i > F_j)$,

        j is the new solution

end

consider the best solution

Rank the obtained solutions and find the current best;

end

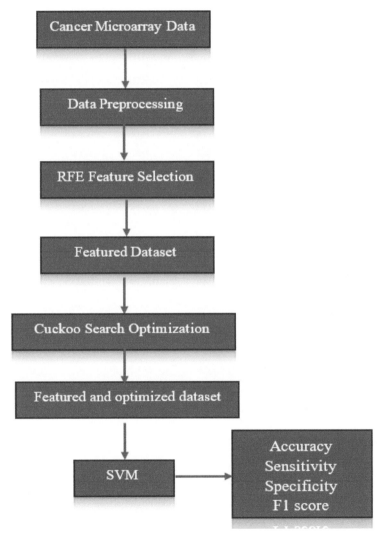

Figure 9.1 Workflow of the proposed work.

Table 9.2   Dimension of the used dataset.

Dataset	Size
Prostate cancer	102×2196
Breast cancer	10×699
Colon cancer	50×2001
Lung cancer	178×1628
Cervical cancer	858×36

Table 9.3   Cross-validation for the proposed work with different cancer datasets.

Number of folds		Specificity	Sensitivity	F1-score	Accuracy
20	Prostate cancer	0.978	0.978	0.977	0.978
	Breast cancer	0.953	0.95	0.951	0.95
	Colon cancer	0.966	0.961	0.962	0.961
	Lung cancer	0.978	0.984	0.979	0.978
	Cervical cancer	0.944	0.945	0.942	0.945
10	Prostate cancer	0.961	0.967	0.966	0.967
	Breast cancer	0.953	0.95	0.951	0.95
	Colon cancer	0.966	0.961	0.962	0.996
	Lung cancer	0.978	0.984	0.979	0.978
	Cervical cancer	0.944	0.945	0.942	0.945
5	Prostate cancer	0.973	0.972	0.971	0.972
	Breast cancer	0.962	0.961	0.962	0.961
	Colon cancer	0.962	0.966	0.967	0.956
	Lung cancer	0.978	0.986	0.979	0.978
	Cervical cancer	0.94	0.939	0.935	0.939

9.4.3 Dataset

Different cancer microarray datasets have been considered for the analysis of the proposed model such as the breast cancer, prostate cancer, cervical cancer, colon cancer, lung cancer, and brain cancer, having different samples and attributes [19]. The attributes matrix is specified in Table 9.2.

9.5 Empirical Analysis

To implement the proposed work, Python 3.0.3 has been taken as the primary environment with Anaconda IDE with a system having i5 processor, 8 GB RAM, and Windows 10 OS. The RFE is considered as the feature selection algorithm, cuckoo search is used as the optimization algorithm, and SVM is used as the classifier. The performance of the proposed system is compared with SVM and RFE SVM to show the effectiveness of the proposed system. For validation, the leave-one-out and cross-validation techniques are used. Tables 9.3 and 9.4 show the cross-validation and LOO validation

Table 9.4 Leave-one-out validation for the proposed work with different cancer datasets.

	Specificity	**Sensitivity**	**F1-score**	**Accuracy**
Prostate cancer	0.956	0.946	0.950	0.977
Breast cancer	0.947	0.947	0.962	0.989
Colon cancer	0.976	0.968	0.982	0.992
Lung cancer	0.960	0.973	0.970	0.988
Cervical cancer	0.969	0.955	0.969	0.979

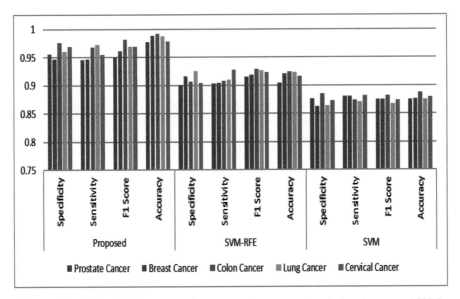

Figure 9.2 LOO validation comparison among the proposed work, SVM-RFE, and SVM.

performance measures. Figure 9.2-9.5 shows performance evaluation of the proposed work.

9.6 Conclusion

We have briefed the concepts of the machine learning microarray data and its importance on cancer classification. Simultaneously, we have summarized the basic difficulties faced during the microarray data analysis. To overcome the well-known issues of the microarray data, the feature selection and feature extract are the possible solutions present. In our research, we have focused on the feature selection approach. We have proposed a model based

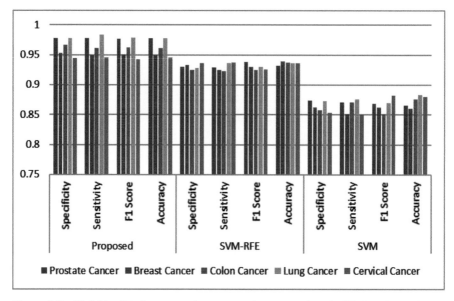

Figure 9.3   20-fold validation comparison among the proposed work, SVM-RFE, and SVM.

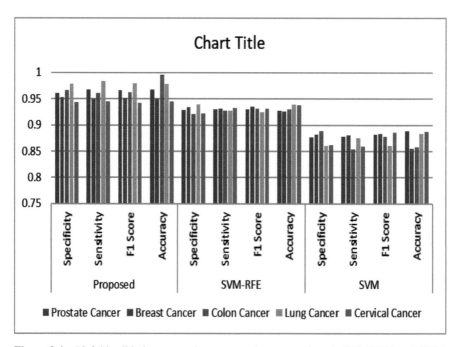

Figure 9.4   10-fold validation comparison among the proposed work, SVM-RFE, and SVM.

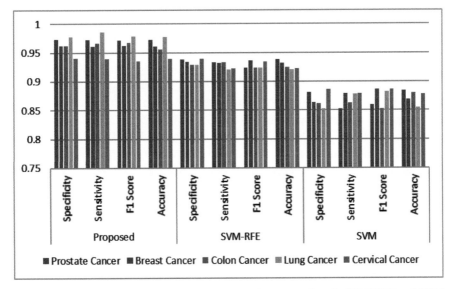

Figure 9.5 Fivefold validation comparison among the proposed work, SVM-RFE, and SVM.

on RFE-CS-SVM, where RFE has been considered as the feature selection algorithm, CS as the optimization technique, and SVM as the classification technique. The leave-one-out and cross-fold validation techniques are considered for the performance evaluation. The empirical analysis shows that the proposed methods show the highest accuracy for colon cancer. The proposed methods show 99.2% accuracy for the leave-one-out validation method. In the case of the cross-fold validation technique, the proposed methods show 99.6% accuracy for colon cancer with the number of folds as 10.

References

[1] Zhu, J., & Hastie, T. (2004). Classification of gene microarrays by penalized logistic regression. Biostatistics, 5(3), 427–443.

[2] Zhou, X., Liu, K. Y., & Wong, S. T. (2004). Cancer classification and prediction using logistic regression with Bayesian gene selection. Journal of Biomedical Informatics, 37(4), 249–259.

[3] Zhao, H., Qi, S., & Dong, Q. (2012, October). Predicting prostate cancer progression with penalized logistic regression model based on co-expressed genes. In 2012 5th International Conference on BioMedical Engineering and Informatics (pp. 976–980). IEEE.

[4] Morais-Rodrigues, F., Silvério-Machado, R., Kato, R. B., Rodrigues, D. L. N., Valdez-Baez, J., Fonseca, V., ... & Dos Santos, M. A. (2020). Analysis of the microarray gene expression for breast cancer progression after the application modified logistic regression. Gene, 726, 144168.

[5] Fan, L., Poh, K. L., & Zhou, P. (2009). A sequential feature extraction approach for naïve bayes classification of microarray data. Expert Systems with Applications, 36(6), 9919–9923.

[6] Wu, M. Y., Dai, D. Q., Shi, Y., Yan, H., & Zhang, X. F. (2012). Biomarker identification and cancer classification based on microarray data using laplace naive bayes model with mean shrinkage. IEEE/ACM transactions on computational biology and bioinformatics, 9(6), 1649–1662.

[7] Sahu, B., & Mohanty, S. N. (2021). CMBA-SVM: a clinical approach for Parkinson disease diagnosis. International Journal of Information Technology, 13(2), 647–655.

[8] Sahu, B., & Panigrahi, A. (2020, February). Efficient role of machine learning classifiers in the prediction and detection of breast cancer. In 5th International Conference on Next Generation Computing Technologies (NGCT-2019).

[9] Ayyad, S. M., Saleh, A. I., & Labib, L. M. (2019). Gene expression cancer classification using modified K-Nearest Neighbors technique. Biosystems, 176, 41–51.

[10] Mahfouz, M. A., Shoukry, A., & Ismail, M. A. (2021). EKNN: Ensemble classifier incorporating connectivity and density into kNN with application to cancer diagnosis. Artificial Intelligence in Medicine, 111, 101985.

[11] Sahu, B., Dash, S., Mohanty, S. N., & Rout, S. K. (2018). Ensemble comparative study for diagnosis of breast cancer datasets. International Journal of Engineering & Technology, 7(4.15), 281–285.

[12] Huo, Y., Xin, L., Kang, C., Wang, M., Ma, Q., & Yu, B. (2020). SGL-SVM: a novel method for tumor classification via support vector machine with sparse group Lasso. Journal of theoretical biology, 486, 110098.

[13] Sahu, B., Gouse, M., Pattnaik, C. R., & Mohanty, S. N. (2021). MMFA-SVM: New bio-marker gene discovery algorithms for cancer gene expression. Materials Today: Proceedings.

[14] Albashish, D., Hammouri, A. I., Braik, M., Atwan, J., & Sahran, S. (2021). Binary biogeography-based optimization based SVM-RFE for feature selection. Applied Soft Computing, 101, 107026.

[15] Mabarti, I. (2020). Implementation of Minimum Redundancy Maximum Relevance (MRMR) and Genetic Algorithm (GA) for Microarray Data Classification with C4. 5 Decision Tree. Journal of Data Science and Its Applications, 3(1), 38–47.

[16] Ghiasi, M. M., & Zendehboudi, S. (2021). Application of decision tree-based ensemble learning in the classification of breast cancer. Computers in Biology and Medicine, 128, 104089.

[17] Ram, M., Najafi, A., & Shakeri, M. T. (2017). Classification and biomarker genes selection for cancer gene expression data using random forest. Iranian journal of pathology, 12(4), 339.

[18] Abdulla, M., & Khasawneh, M. T. (2020). G-Forest: An ensemble method for cost-sensitive feature selection in gene expression microarrays. Artificial Intelligence in Medicine, 108, 101941.

[19] Wang, S., Wang, Y., Wang, D., Yin, Y., Wang, Y., & Jin, Y. (2020). An improved random forest-based rule extraction method for breast cancer diagnosis. Applied Soft Computing, 86, 105941.

[20] Segera, D., Mbuthia, M., & Nyete, A. (2019). Particle swarm optimized hybrid kernel-based multiclass support vector machine for microarray cancer data analysis. BioMed research international, 2019.

[21] Mert, A., Kilic, N., & Akan, A. (2011, September). Breast cancer classification by using support vector machines with reduced dimension. In Proceedings ELMAR-2011 (pp. 37–40). IEEE.

[22] Pashaei, E., Pashaei, E., & Aydin, N. (2019). Gene selection using hybrid binary black hole algorithm and modified binary particle swarm optimization. Genomics, 111(4), 669–686.

[23] Alshamlan, H., Badr, G., & Alohali, Y. (2015). mRMR-ABC: a hybrid gene selection algorithm for cancer classification using microarray gene expression profiling. Biomed research international, 2015.

[24] Jain, D., & Singh, V. (2020). A novel hybrid approach for chronic disease classification. International Journal of Healthcare Information Systems and Informatics (IJHISI), 15(1), 1–19.

[25] Gupta, K., & Janghel, R. R. (2019). Dimensionality reduction-based breast cancer classification using machine learning. In *Computational Intelligence: Theories, Applications and Future Directions-Volume I* (pp. 133–146). Springer, Singapore.

[26] Jain, I., Jain, V. K., & Jain, R. (2018). Correlation feature selection based improved-binary particle swarm optimization for gene selection and cancer classification. Applied Soft Computing, 62, 203-215.

[27] Mabarti, I. (2020). Implementation of Minimum Redundancy Maximum Relevance (MRMR) and Genetic Algorithm (GA) for Microarray Data Classification with C4. 5 Decision Tree. *Journal of Data Science and Its Applications*, 3(1), 38–47.

[28] Mohamed, N. S., Zainudin, S., & Othman, Z. A. (2017). Metaheuristic approach for an enhanced mRMR filter method for classification using

drug response microarray data. *Expert Systems with Applications*, *90*, 224–231.

[29] Astuti, W. (2018, March). Support vector machine and principal component analysis for microarray data classification. In *Journal of Physics: Conference Series* (Vol. 971, No. 1, p. 012003). IOP Publishing.

[30] Adiwijaya, W. U., Lisnawati, E., Aditsania, A., & Kusumo, D. S. (2018). Dimensionality reduction using principal component analysis for cancer detection based on microarray data classification. *Journal of Computer Science*, *14*(11), 1521–1530.

[31] Pati, A., Parhi, M., & Pattanayak, B. K. (2021). COVID-19 Pandemic Analysis and Prediction Using Machine Learning Approaches in India. In Advances in Intelligent Computing and Communication (pp. 307–316). Springer, Singapor

[32] Sahu, B., Panigrahi, A., Sukla, S., & Biswal, B. B. (2020). MRMR-BAT-HS: a clinical decision support system for cancer diagnosis. Leukemia, 7129(73), 48.

[33] Pati, A., Parhi, M., & Pattanayak, B. K. (2021, January). IDMS: An Integrated Decision Making System for Heart Disease Prediction. In 2021 1st Odisha International Conference on Electrical Power Engineering, Communication and Computing Technology (ODICON) (pp. 1–6). IEEE.

[34] Sahu, B., Panigrahi, A., Mohanty, S., & Sobhan, S. (2020). A hybrid cancer classification based on SVM optimized by PSO and reverse firefly algorithm. International Journal of Control and Automation, 13(4), 506–517.

[35] Sahu, B., Panigrahi, A., Pani, S., Swagatika, S., Singh, D., & Kumar, S. (2020, July). A crow particle swarm optimization algorithm with deep neural network (CPSO-DNN) for high dimensional data analysis. In 2020 International Conference on Communication and Signal Processing (ICCSP) (pp. 0357–0362). IEEE.

[36] Sahu, B., Badajena, J. C., Panigrahi, A., Rout, C., & Sethi, S. (2020). An Intelligence-Based Health Biomarker Identification System Using Microarray Analysis. In Applied intelligent decision making in machine learning (pp. 137–161). CRC Press.

[37] Sahu, B., Panigrahi, A., & Rout, S. K. (2020). DCNN-SVM: A New Approach for Lung cancer Detection. In Recent Advances in Computer-Based Systems, Processes, and Applications (pp. 97–105). CRC Press

10

Feature Selection/Dimensionality Reduction

Divya Stephen[1], S. U. Aswathy[2], Priya P. Sajan[3], and Jyothi Thomas[4]

[1]Department of Computer Science and Engineering, Jyothi Engineering College, India.
[2]Department of Computer Science and Engineering, Marian Engineering College, India.
[3]C-DAC, India.
[4]Department of Computer Science and Engineering, Christ University, India.
Email: divyasrdj@gmail.com; aswathy.su@gmail.com;
priyasajans@gmail.com; j.thomas@christuniversity.in

Abstract

In today's world, medical image analysis is a critical component of research, and it has been extensively explored over the last few decades. Machine learning in healthcare is a fantastic advancement that will improve disease detection efficiency and accuracy. In many circumstances, it will also allow for early detection and treatment in remote or developing areas. The amount of medical data created by various applications is growing all the time, creating a bottleneck for analysis and necessitating the use of a machine learning method for feature selection and dimensionality reduction techniques. Feature selection is an important concept of machine learning since it affects the model's performance and the data parameters you utilize to train your machine learning models to have a big influence on the performance. The approach of minimizing the number of inputs in training data by reducing the dimension of your feature set is known as dimensionality reduction. Reduced dimensionality aids in the overall performance of the machine learning algorithms.

10.1 Introduction

The mortality rate from numerous diseases is steadily rising. A quick diagnosis at an early stage can help with the implementation of effective sickness preventive and mitigation strategies, as well as a better prognosis. Machine learning's benefit in healthcare is its ability to analyze enormous datasets beyond human capability and then consistently turn data analysis into clinical insights that aid clinicians in planning and providing care, resulting in better outcomes and lower healthcare costs. This chapter introduces the most recent feature selection as well as the dimensionality reduction approaches discovered and/or applied in disease diagnosis using machine learning techniques. The selection of appropriate attributes is necessary to enhance the accuracy of the system. Reduced dimensionality aids in the overall performance of machine learning algorithms.

A feature is defined as an "individual measurable attribute or a typical aspect of an event under observation" in machine learning. Each feature provides a specific piece of information that aids in analysis. The importance of feature selection in the development of accurate machine learning models cannot be overstated. Let us say that the machine learning model for the provided dataset learns the mapping between the input features and the target variable. As a result, the model can reliably predict the target variable in a new dataset when the target variable is unknown. Many factors influence the performance of a machine learning model, including:

- Algorithm selection

- The model's features that were used to train it

- Input data quality

- Parameters used in the algorithm

Occasionally, the set of features in their raw form in a dataset does not provide the best information for training and prediction. As a result, it is advantageous to remove the incompatible and superfluous features from our dataset. This manner reduces bodily intervention in fact analysis and makes the feature interpretation easy and equipped to apply. The technique allows us to pick out the most centered variable correlating with other variables. Hence, the version performance is expanded with the selected features.

First of all, let us know the difference between the feature selection and the dimensionality reduction; then we will go into detail. Often, feature selection and dimensionality reduction are grouped together. The two methodologies for reducing the number of characteristics in a dataset have substantial differences. Feature selection is just picking and choosing which features to include and exclude without changing them; through weightage computation, these algorithms assist us in identifying the most relevant features, whereas dimensionality reduction reduces the dimensions of characteristics.

10.2 Feature Selection

Feature selection is defined as "the process of picking a subset of relevant characteristics for use in model creation," or "the selection of the most important features." In pattern recognition and image processing, feature extraction is a sort of dimensionality reduction. When an algorithm's input data becomes too huge to handle and is suspected of being notoriously redundant, it is translated into a smaller representation set of features (feature vector). Feature selection as a dimensionality reduction approach aims to pick a small segment of the salient set of features by deleting unwanted, redundant, or chaotic characteristics. Feature selection is crucial for efficient learning performance, improved learning precision, reduced processing costs, and improved model interpretability.

Now we will look at the feature selection process, which gives you a thorough and organized review of feature selection kinds, procedures, and techniques from both a data and algorithmic aspect. Feature engineering, which includes feature extraction and feature selection and is a fundamental building element of modern machine learning pipelines, is essential to all machine learning workflows. In spite of the fact that feature extraction and feature selection approaches are comparable in certain ways, they are frequently interchanged. Feature extraction is the process of extracting additional variables from raw data using domain expertise to make machine learning algorithms work. The most consistent, useful, and non-redundant features are chosen during the feature selection process [1]. The major stages in the feature selection process are depicted in Figure 10.1, and the illustration is shown in Figure 10.2.

Feature selection strategies include the following goals:

- researchers and users will find it easier to interpret models that have been simplified;

- training durations that are shorter;

- keeping the curse of dimensionality at bay;

- improved generalization by eliminating overfitting (formally, reduction of variance).

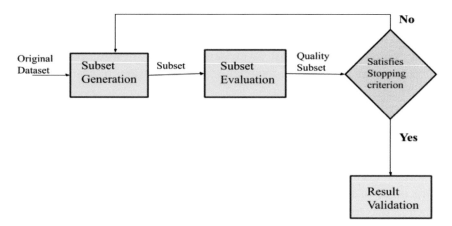

Figure 10.1   Major stages in the feature selection process.

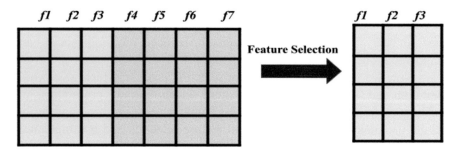

Figure 10.2   Feature selection illustration.

Because of the wide range of established analytic methods available to researchers, dataset size reduction is becoming more important, even while the typical dataset grows in terms of the number of characteristics and samples. The goal of this technique is to increase relevance while minimizing repetition. To accomplish effective data compression, feature selection algorithms can be utilized in pre-processing. This aids in the discovery of precise data models. A number of search approaches have been suggested since an exhaustive search for an ideal feature subset is infeasible in most circumstances. In classification, grouping, and regression tasks, feature selection is often utilized.

10.2.1 Characteristics

A feature selection algorithm's goal is to find relevant characteristics based on a relevance criterion. However, in machine learning, the concept of relevance has yet to be thoroughly defined and agreed upon. The concept of being relevant in relation to an objective is one of the most basic definitions of relevance. There are various factors to consider when describing feature selection algorithms. In the context of the present study, this characterization can be described as a hypothesis space search problem.

- Search organization: A broad technique for exploring the hypothesis space.

- Generation of successors: The method of proposing alternative versions (successor candidates) of the existing hypothesis.

- Evaluation measure: A tool for evaluating successor candidates, which allows you to compare different assumptions to help direct your search.

10.2.2 Classification of feature selection methods

There are supervised and unsupervised approaches for selecting relevant features by decreasing redundant and irrelevant feature selection, resulting in more accuracy. Supervised feature selection refers to the method that uses the output label class for feature selection. They use the target variables to identify the variables that can increase the efficiency of the model, and unsupervised feature selection refers to the method that does not need the output label class for feature selection. We use them for unlabeled data. Different methodologies might be used depending on how and when the utility of specific qualities is assessed, like filter, wrapper, embedded, and hybrid methods, which are also known as the evaluation methods in feature selection [9]. Another type of evaluation method, ensemble feature selection, has been developed in recent years [8].

Figure 10.3 shows the taxonomy for feature selection.

10.2.2.1 Supervised feature selection methods

We have filter, wrapper, hybrid, and embedded methods in supervised feature selection based on how they interact with the learning model [20].

- **Filter methodology:**
  Figure 10.4 depicts the Filter based model. Statistical metrics are used to choose features in the filter technique. It is unaffected by the learning method and takes less time to compute. Information gain, Fisher score, chi-square test, correlation coefficient, and variance threshold are some of the statistical measurements used to understand the worth of the features. The filter approach identifies irrelevant attributes and filters

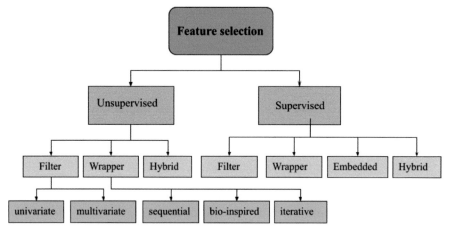

Figure 10.3 Taxonomy for feature selection.

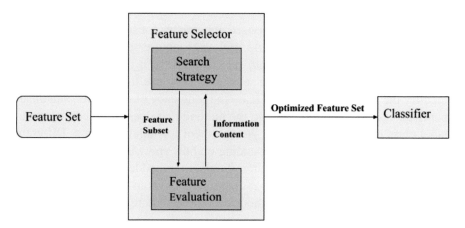

Figure 10.4 Filter-based feature selection.

away redundant columns from the models using the selected measure. It allows you to isolate specific measures that enrich a model, and following the computation of feature scores, the columns are ranked.

- **Wrapper methodology:**
  This methodology treats set selection as a search problem, with multiple combinations being constructed, assessed, and evaluated by comparing to one another. To evaluate a set of features and calculate model performance ratings, a predictive model is utilized. The wrapper technique's performance is determined by the classifier, and the optimal subset of attributes is chosen depending on the classifier's findings. Wrapper based model is shown in Figure 10.5.

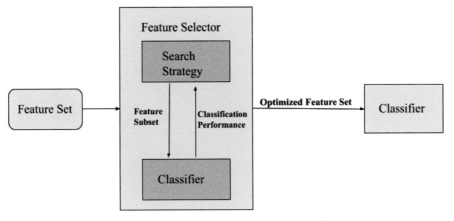

Figure 10.5 Wrapper-based feature selection.

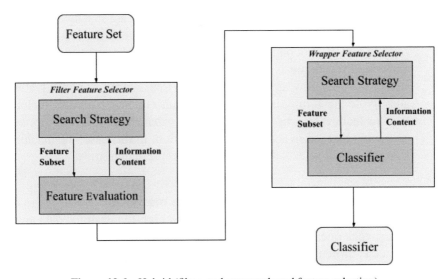

Figure 10.6 Hybrid (filter- and wrapper-based feature selection).

- **Hybrid methodology:**
  In hybrid, what you choose to mix determines how you create hybrid feature selection algorithms. The most important thing is to choose the methods you will utilize and follow its procedures. In the first phase, to produce a feature ranking list, we make use of the ranking methods and then conduct wrapper methods on the top k features from that list. As a result, we can use these filter-based rangers to shrink the feature space of our dataset, lowering the time complexity of the wrapper techniques. Figure 10.6 shows the hybrid model.

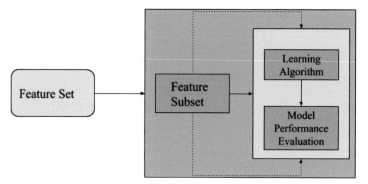

Figure 10.7   Embedded or intrinsic-based feature selection.

- **Embedded or intrinsic methodology:**
  Embedded or intrinsic feature selection methods are machine learning models that have feature selection built-in as part of the learning process. Figure 10.7 depicts the embedded model. The most widely used embedded technique is the decision tree algorithm. In each iterative step of the tree-growing process, decision tree algorithms pick a feature and divide the sample set into smaller subsets. Some models include built-in feature selection, which indicates that the model incorporates predictors that aid in accuracy maximization. In this case, the machine learning model selects the optimum data representation.

10.2.2.2 Unsupervised feature selection methods

Unsupervised approaches are frequently used in high-dimensional data analysis due to a lack of easily available labels [3]. Determining the right number of attributes is difficult. Filter, wrapper, and hybrid approaches are unsupervised feature selection methods that are classified according to how they interact with the learning model.

- **Filter methodology:**
  Unsupervised feature selection strategies based on the filter mechanism are divided into two categories: univariate and multivariate.

  Univariate approaches, often known as ranking-based unsupervised feature selection methods, evaluate each feature according to a set of criteria, resulting in an ordered ranking list of characteristics from which the final subset is selected. These approaches are good at finding and deleting unneeded features, but they cannot eliminate duplicate features because they do not take into consideration possible feature dependencies.

Multivariate filtering, on the other hand, considers the significance of traits collectively rather than individually. Multivariate approaches can handle features that are repetitious or meaningless. As a result, learning algorithms that use the subset of features provided by multivariate approaches generally outperform univariate methods in terms of accuracy.

- **Wrapper methodology:**
  There are three major types, according to the feature search strategy: sequential, bio-inspired, and iterative. The sequential process is used to add or remove features in a gradual manner. The implementation of sequential search algorithms is easy and quick. By introducing unpredictability into the search process, the bio-inspired method seeks to prevent local optima. Iterative approaches avoid a systematic search by recasting the unsupervised feature selection problem as an estimation problem. Wrapper approaches use the results of a clustering algorithm to evaluate selected features. The approaches used in this methodology are notable for identifying feature subsets that aid in increasing the quality of the clustering algorithm's chosen outputs. Wrapper techniques, on the other hand, have the disadvantage of having a high processing cost and being restricted to use with a single clustering algorithm.

- **Hybrid methodology:**
  In order to attain a fair mix of efficiency (computational effort) and efficacy, hybrid methods aim to combine the benefits of both filter and wrapper approaches. Hybrid methods involve a filter phase in which the features are ordered or picked by imposing a measure depending on the intrinsic attributes of the data in order to use the filter and wrapper approaches. At the wrapper stage, specific feature subsets are examined using a clustering approach in order to find the best one. Hybrid methods are divided into two categories: ranking-based methods and non-ranking-based approaches.

Figure 10.8 gives an overview of the filter, wrapper, and embedded methods.

10.2.3 Importance of feature selection in machine learning

Machine learning is based on a basic principle: if you put garbage in, garbage will come out. When I say garbage, I am referring to data noise. When

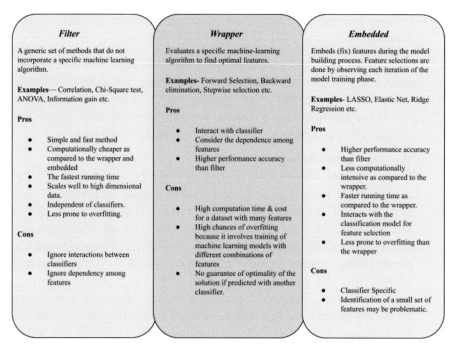

Figure 10.8   Overview of filter, wrapper, and embedded methods.

there are a lot of features, this becomes much more significant. When constructing an algorithm, you do not have to employ every feature available to you. You may help your algorithm by feeding only the most important features into it.

The following are the most compelling reasons to employ feature selection:

- It allows the machine learning system to learn more quickly.

- It simplifies a model's complexity and makes it easier to understand.

- When the proper subset is picked, a model's accuracy improves.

- Overfitting is reduced.

10.3 Dimensionality Reduction

The number of parameters or features in the source determines the dimensionality of a dataset. Strategies for minimizing the input variables are referred to as dimensionality reduction. The dimension of a dataset refers to the number of input variables and attributes. A data model becomes more complex to model as more input features are included. Reducing the complexity of a dataset can help train a prediction model in machine learning. This portion will provide you a gentle exposure to dimensionality reduction for machine learning.

Dimensionality reduction refers to methods for reducing the number of input parameters in the training phase. When working with high-dimensional input, it is frequently essential to minimize dimensionality by projecting the data to a lower-dimensional subspace that preserves the data's significance. Dimensionality reduction is the term for this process [3]. The presence of millions and millions of input parameters is referred to as high-dimensionality. Reduced computational dimensions can mean fewer variables or a better machine learning model framework, which are referred to as degrees-of-freedom. Overfitting the training data is more likely with a model with too many degrees of freedom, leading to poor performance when confronted with new data. Prior to modeling, dimensionality reduction is a data preparation tool used on data. Reduced dimensionality results in a more compact, easy-to-understand representation of the goal notion, allowing the users to focus on the most important aspects [5]. As a result, any dimensionality reduction done on the training set must also be done on other data, such as a test set, validation dataset, or information needed to assume using the final model.

10.3.1 Techniques for dimensionality reduction

The most often used data-dimensionality reduction approaches include the following:

- Ratio of missing values: We utilize this strategy to make the number of variables smaller in a set of data if there are too many missing values. We can eliminate variables with a wide range of missing values. As a result, record columns with a missing value ratio greater than a certain threshold may be eliminated. The decrease is more aggressive the higher the edge.

- Low variance: This method is used to select and remove constant variables from the dataset. Variables with minimal variance do not have an undue impact on the target variable, and so these variables can be successfully deleted. Because the variance is dependent on the number of columns, normalization is required before applying this strategy.

- High correlation filter: Dimensions with a higher correlation can reduce a model's overall performance. Furthermore, having more than one variable with equivalent information is not necessarily true. The Pearson correlation matrix has been used to find variables that have a high level of correlation. And then pick one based on VIF (variance inflation factor) consumption. Higher-valued variables (VIF > 5) may be eliminated.

- Random forest: It is a decision-tree-like technique that can be used to choose characteristics and classify them. To reduce dimensionality, one way is to build a large, well-planned collection of trees against an intended feature, and then using usage statistics for each variable to identify the most significant subset of characteristics. This is one of the most often used strategies for determining the significance of each feature in a dataset.

- Principal component analysis (PCA): It is the most widely used linear data modification technique that divides the data into several components to explain as many variations as possible. PCA is a statistical method for orthogonally transforming a dataset's initial numeric dimensions into a new collection of numeric dimensions called principal components.

- Backward feature elimination: Here, the given classifying algorithm is trained on m input columns at each cycle. Then, one by one, each input column is removed, and the same model is trained on $m - 1$ columns. We remove the input column with the smallest error rate increase, leaving $m - 1$ input columns. The classification is then carried out with $m - 2$ columns, and so on. Every cycle k generates a model with $m - k$ training columns and a false positive rate of $e. (k)$. We now use the highest acceptable false positive rate to determine the lowest number of columns needed to achieve that classification effectiveness using the machine learning technique.

- Forward feature construction: Backward feature removal is the inverse of this technique. We start by attaching a single column at a time, starting with the one that results in the most performance gain. Backward feature deletion and forward feature building are both time- and computational-intensive algorithms. They are only beneficial if you start with a dataset with a modest number of input columns. Because both backward feature elimination and forward feature construction take a long time to compute, they are often utilized on smaller datasets.

- Factor analysis: This method is best used in circumstances when a large number of variables are highly connected. It separates the variables into

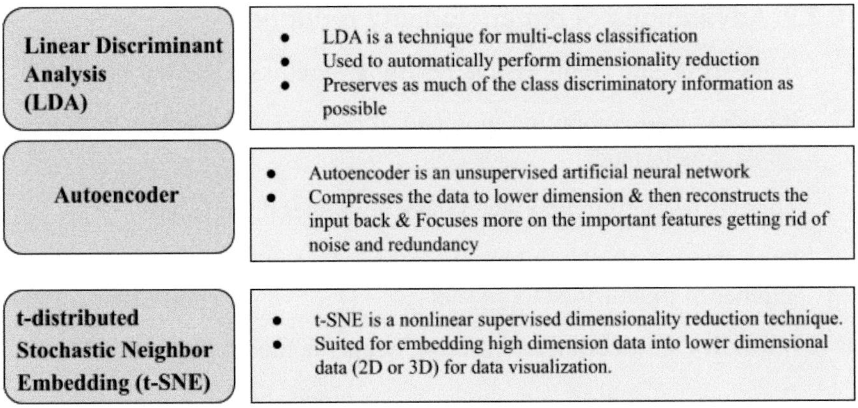

Linear Discriminant Analysis (LDA)	• LDA is a technique for multi-class classification • Used to automatically perform dimensionality reduction • Preserves as much of the class discriminatory information as possible
Autoencoder	• Autoencoder is an unsupervised artificial neural network • Compresses the data to lower dimension & then reconstructs the input back & Focuses more on the important features getting rid of noise and redundancy
t-distributed Stochastic Neighbor Embedding (t-SNE)	• t-SNE is a nonlinear supervised dimensionality reduction technique. • Suited for embedding high dimension data into lower dimensional data (2D or 3D) for data visualization.

Figure 10.9 LDA, autoencoder, and *t*-SNE.

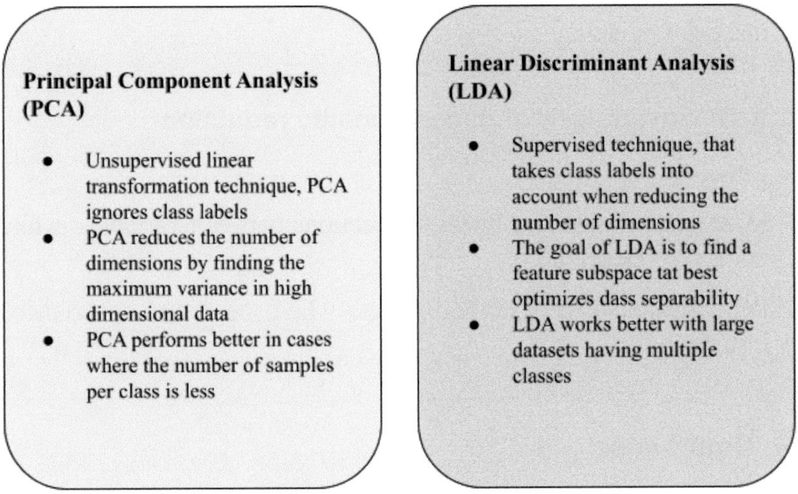

Principal Component Analysis (PCA)

- Unsupervised linear transformation technique, PCA ignores class labels
- PCA reduces the number of dimensions by finding the maximum variance in high dimensional data
- PCA performs better in cases where the number of samples per class is less

Linear Discriminant Analysis (LDA)

- Supervised technique, that takes class labels into account when reducing the number of dimensions
- The goal of LDA is to find a feature subspace tat best optimizes dass separability
- LDA works better with large datasets having multiple classes

Figure 10.10 PCA vs. LDA.

groups based on their association and assigns a factor to each category. These characteristics, however, are difficult to observe. Factor analysis can be carried out in one of the following two ways:

○ EFA (exploratory factor analysis);

○ CFA (confirmatory factor analysis).

• We also have the three new techniques shown in Figure 10.9.

From all these techniques, we mainly deal with PCA and LDA strategies; a short description is given in Figure 10.10.

10.3.2 Advantages of dimensionality reduction

- It assists in data compression, resulting in reduced storage space.

- It has reduced computing time and, if necessary, the removes unnecessary characteristics.

- There is less computing when there are fewer dimensions.

- Dimensions can also allow algorithms that are unsuitable for a high number of dimensions to be used.

- It handles multicollinearity, which helps the model perform better.

- Data can be plotted and shown more precisely if the dimensions are reduced to 2D or 3D.

- It also aids in the reduction of noise, allowing to increase the performance of models.

10.3.3 Disadvantages of dimensionality reduction

- Data loss.

- PCA tends to uncover linear connections between variables, which is not always desired.

- PCA also fails to describe datasets when mean and covariance are insufficient.

10.4 Conclusion

Feature selection is a frequent optimization problem in which the best answer requires an extensive search, based on a set of criteria. Researchers continue to apply the polynomial time complexity heuristic method for high-dimensional problems, whereas dimensionality reduction approaches aim to reduce the number of input parameters to a predictive model. The difference between feature selection and dimensionality reduction is that feature selection chooses which features to preserve or eliminate from the dataset, whereas dimensionality reduction projects the data to generate entirely new input features. As a result, rather than being an alternative to feature selection, dimensionality reduction is a type of feature selection. In dimensionality reduction, there is no single best dimensionality reduction strategy, and there is no technique-to-problem mapping. Instead, use systematic controlled testing to determine which dimensionality reduction techniques, when combined with your preferred model, offer the greatest results on your dataset. All input features are assumed to have the same scale or distribution in linear algebra and manifold learning approaches. This means that if the input variables have multiple scales or units, normalizing or standardizing the data before using these methods is a good idea.

References

[1] Yanfang Liu, Dongyi Ye, Wenbin Li, Huihui Wang, Yang Gao, Robust neighborhood embedding for unsupervised feature selection, Knowledge-Based Systems, Volume 193, 2020, 105462, ISSN 0950-7051, https://doi.org/10.1016/j.knosys.2019.105462.

[2] José Jair A. Mendes Junior, Melissa L.B. Freitas, Hugo V. Siqueira, André E. Lazzaretti, Sergio F. Pichorim, Sergio L. Stevan, Feature selection and dimensionality reduction: An extensive comparison in hand gesture classification by sEMG in eight channels armband approach, Biomedical Signal Processing and Control, Volume 59, 2020, 101920, ISSN 1746-8094, https://doi.org/10.1016/j.bspc.2020.101920.

[3] Rizgar R. Zebari, Adnan Mohsin Abdulazeez, Diyar Qader Zeebaree, Dilovan Asaad Zebari, Jwan Najeeb Saeed, A Comprehensive Review of Dimensionality Reduction Techniques for Feature Selection and Feature Extraction, JOURNAL OF APPLIED SCIENCE AND TECHNOLOGY TRENDS, Vol. 01, No. 02, pp. 56 –70 (2020) ISSN: 2708-0757, doi: 10.38094/jastt1224

[4] M. El-Hasnony, S. I. Barakat, M. Elhoseny and R. R. Mostafa, "Improved Feature Selection Model for Big Data Analytics," in *IEEE Access*, vol. 8, pp. 66989-67004, 2020, doi: 10.1109/ACCESS.2020.2986232.

[5] G. Taşkın, H. Kaya and L. Bruzzone, "Feature Selection Based on High Dimensional Model Representation for Hyperspectral Images," in IEEE Transactions on Image Processing, vol. 26, no. 6, pp. 2918-2928, June 2017, doi: 10.1109/TIP.2017.2687128.

[6] Beatriz Remeseiro, Veronica Bolon-Canedo, A review of feature selection methods in medical applications, Computers in Biology and Medicine, Volume 112, 2019, 103375, ISSN 0010-4825, https://doi.org/10.1016/j.compbiomed.2019.103375.

[7] Mengmeng Li, Haofeng Wang, Lifang Yang, You Liang, Zhigang Shang, Hong Wan, Fast hybrid dimensionality reduction method for classification based on feature selection and grouped feature extraction, Expert Systems with Applications, Volume 150, 2020, 113277,

ISSN 0957-4174, https://doi.org/10.1016/j.eswa.2020.113277. (https://www.sciencedirect.com/science/article/pii/S0957417420301020)

[8] C. Lazar, J. Taminau, S. Meganck, D. Steenhoff, A. Coletta, C. Molter, V. de Schaetzen, R. Duque, H. Bersini, and A. Nowe, "A Survey on Filter Techniques for Feature Selection in Gene Expression Microarray Analysis," IEEEACM Trans Comput Biol Bioinforma., vol. 9, no. 4, pp. 1106–1119, Jul. 2012

[9] Y. Leung and Y. Hung, "A Multiple-Filter-Multiple-Wrapper Approach to Gene Selection and Microarray Data Classification," IEEEACM Trans Comput Biol Bioinform., vol. 7, no. 1, pp. 108–117, Jan. 2010.

[10] A. K. Shafreen Banu, S. Hari Ganesh, Ph.D., A Study 0/ Feature Selection Approaches/or Classification, IEEE Sponsored 2nd International Conference on Innovations in Information Embedded and Communication Systems ICIIECS'15

[11] J. Doak, An Evaluation of Feature Selection Methods and Their Application to Computer Security. University of California, Computer Science, 1992.

[12] S.Visalakshi V. Radha, A Literature Review of Feature Selection Techniques and Applications Review of feature selection in data mining, 978-1-4799-3975-6/14/$31.00 ©2014 IEEE

[13] Jie Cai, Jiawei Luo, Shulin Wang, Sheng Yang, Feature selection in machine learning: A new perspective, Neurocomputing 300 (2018) 70–79, https://doi.org/10.1016/j.neucom.2017.11.077 0925-2312/© 2018 Elsevier B.V.

[14] S. Van Landeghem, T. Abeel, Y. Saeys, Y. Van de Peer, Discriminative and informative features for biomolecular text mining with ensemble feature selection, Bioinformatics 26 (2010) 554–560.

[15] H. Liu, J. Li, L. Wong, A comparative study on feature selection and classification methods using gene expression profiles and proteomic patterns, Genome Inform. 13 (2002) 51–60.

[16] G. Li, X. Hu, X. Shen, X. Chen, Z. Li, A novel unsupervised feature selection method for bioinformatics data sets through feature clustering, in Proceedings of IEEE International Conference on Granular Computing, 2008, pp. 41–47.

[17] Jianyu Miao a,c, Lingfeng Niu b, A Survey on Feature Selection, Information Technology and Quantitative Management (ITQM 2016), © by Elsevier B.V (http://creativecommons.org/licenses/by-nc-nd/4.0/).

[18] Zhaleh Manbari, Fardin AkhlaghianTab,Chiman Salavati. "Hybrid fast unsupervised feature selection for high-dimensional data", Expert Systems with Applications, 2019

[19] Amit Singh & Abhishek Tiwari, 2021. "A Study of Feature Selection and Dimensionality Reduction Methods for Classification-Based Phishing Detection System," International Journal of Information Retrieval Research (IJIRR), IGI Global, vol. 11(1), pages 1-35, January.

[20] Ahmed Hashem El Fiky, Ayman Elshenawy, Mohamed Ashraf Madkour. "Detection of Android Malware using Machine Learning", 2021 International Mobile, Intelligent, and Ubiquitous Computing Conference (MIUCC), 2021

[21] Mohammad Kazem Ebrahimpour, Masoumeh Zare, Mahdi Eftekhari, Gholamreza Aghamolaei. "Occam's razor in dimension reduction: Using reduced row Echelon form for finding linear independent features in high dimensional microarray datasets", Engineering Applications of Artificial Intelligence, 2017

[22] Anupama C V, Nisha Puthiyedth, Neenu R. "Feature Selection Methods for SNP Analysis", 2019 2nd International Conference on Intelligent Computing, Instrumentation and Control Technologies (ICICICT), 2019

11

Information Retrieval using Set-based Model Methods, Tools, and Applications in Medical Data Analysis

S. Castro[1], P. Meena Kumari[2], S. Muthumari[3], and J. Suganthi[4]

[1]Department of Information Technology, Karpagam College of Engineering, India.
[2]Department of Computer Science and Engineering, AVN Institute of Engineering and Technology, India.
[3]Department of Computer Science & Information Technology, S.S. Duraisamy Nadar Mariammal College, India.
[4]Department of Information Science and Engineering, T. John Institute of Technology, India.
Email: suseelcastro@gmail.com; parigimeenakumari@gmail.com; muthu0903@gmail.com; jsuganthi@tjohngroup.com

Abstract

Text analysis plays a significant role in information retrieval (IR) and prediction in real-time data processing platforms. There are certain issues in modeling an IR system such as document indexing, query evaluation, and system evaluation. In recent years, researchers have shown more interest toward the document indexing as the textual data has increased a lot such as social media, healthcare diagnosis, etc. To analyze and investigate the document indexing process, this chapter processes a set-based model with special reference to medical data. The main objective of this set-based model is to find the term weights for index terms for contextual medical term representation. The function of this model is to compute the similarity between a document and a query. Association rule-based evaluation provides the significance of the terms that exist in a document, which is adopted for classification. This seems to be time efficient and improves the mechanism of data retrieval and analysis. The set-based model improves the average precision of the answer

187

set. In terms of computing performance, this model is also competitive and has more parametric executions. Also, this model can be easily understood and interpreted, and it suits real-time processing platforms. This chapter covers a set-based model for reinforcement learning design, a case study, the rank computation process, tools for evaluation, and their applications toward medical data analysis.

11.1 Introduction

In information retrieval (IR), the vector space model (VSM) has been considered to be the best model in proving answers, which humans find similar to search queries. Usually, weighted vectors are used to represent the documents in the VSM and the document that is most relevant to the user query is provided. In the VSM, the similarities of a document to the query are computed by the weights provided to the terms, called index terms, which are assigned to each document. Salton and his team made this vector space model a huge success [11]. There are many other ways to compute the weight of a document and the challenge that we face continuously is finding the best weight terms. The most used schemes here are calculating the time of index term occurrences in the document and calculating the index terms occurring in the total number of documents. Therefore, the main terms used are term frequency (TF) and inverse term frequency (IDF).

The domain of medical data has paved its pathway to other domains of nature such as telemedicine, self-diagnosis, education, etc. Analyzing the medical data and providing its outcome in a better way need some additional precaution measures. There exist various types of terms associated with semi-structured and unstructured medical data. However, the formulated medical textual data required a proper annotation from the medical experts for analysis.

Thus, we are aiming to produce a model that can efficiently be used to calculate the term weights based on the concept called a set theory. Hence, the model has been named as set-based model based on the concept of set theory and it calculates the index terms' weights by using the association rule theory. Association rule theory is algorithmically very interesting and it is very efficient. It also provides a parameterized approach in computing. It is also used to find term co-occurrence factors in the documents. The vector space model is mentioned earlier because the set-based model is the representation of set theory with a vectorial ranking.

To improve the results, mutual dependencies are used among the index terms. In the set-based model, the principle of relational model is used for handling the querying tasks. The relational model is part of the mathematical set theory and it uses cursor in relational databases to provide solutions. Set-based solutions provide T-SQL queries to operate on input tables as a set of rows. The data mining concept is also used in the set-based model. In the set-based model, closed termsets are used to explain the model which is associated with documents. A pair of weights associated with the closed termsets plays an important role in documents and in document collection. Similarly, a pair of weights associated with the closed termsets is essential in query and in query collection. A set of closed terms of algebraic representations for the documents and queries correspond to the 2^t two-dimensional Euclidean space in vector space model where t is the number of unique index terms associated with the documents in the document collections. For the purpose of computing the closed termsets, association rule theory is used. Here, a concept called termsets is used.

Termsets are defined as the term used instead of index terms. Usually, termsets consist of subsets of terms in the collections. We can say that if all the index terms occur in a particular document, then the particular index term occurs in a particular document. In a set-based model, there will be 2^t termsets in which "t" is the size of the vocabulary. Hence, the vocabulary set is the collection of termsets from 1 to 2^t. For long queries, frequent termsets will be useful in reducing the number of termsets required for the set-based model. Subsequently, the set-based model approach provides improved results regarding various collections and it is used to constitute the first information retrieval model that effectively takes the advantages of term dependence with general collections.

11.2 Literature Review

Set-based experiment design for model discrimination using bi-level optimization has been considered as a tedious task in text modeling processes. It is a monotonous job to determine or separate a model that has a non-linear and uncertain condition as parameters. To overcome this task, a new approach, which is called bi-level optimization, has been provided and by using this, the uncertainty models can be easily discriminated based on simple measurement. The bi-level optimization approach is implemented based on the set-based model. In this approach, uncertainty conditions, non-linear models, uncertain parameters, and output models are separated and merged into a feasibility problem. Then, this feasibility problem can be solved efficiently by using validation, estimation, and fault detection and isolation techniques. This approach has two programs. They are outer and inner programs [1].

The outer program is used to determine the input sequence and the inner program is used to separate the output sets. The outer, or otherwise called a no-convex, problem passes the input sequence to the inner program and then the inner program solves the problem of the particular input sequence by using state-of-the-art solvers. Then, it separates the output sets and calculates the distance of the output sets. With this, an input sequence will be found and the outputs are separated by using a single and simple measurement. Finally, the model can be discriminated as valid or invalid by using this bi-level optimization approach. In this task, a design only for the single output sets has been proposed and the output sets should be chosen very carefully [1].

Generally, rough set models are possibly used to extract set-based information in real-time processing platforms. The incomplete information system is substituted with the set-valued information. It is used to analyze a set-based information system for the existing binary relation shortages. Because of the set value's actual semantic, probabilistic equivalence relation is defined by proposing the probabilistic rough set model for the set-valued information system. Based on the probabilistic equivalence relation, set-valued decision information system's attribute reduction is also discussed. In set-valued ordered information system, by analyzing the existing dominance relation defects, the probabilistic dominance relation is proposed. Here, the probabilistic dominance rough set model is used for demonstrating the dominance probability between the object calculations. Pawlak introduced a rough set theory, which is based on a complete information system's assumption. In a set-valued information system, the probabilistic rough set model is based on the concept of set theory, binary relation analysis, probability equivalence relation, and probability rough set model. In the basic concepts, limited torrent relation is discussed [2].

In binary relation analysis, three classification methods are defined and these three classification methods have equal irrationality. In probabilistic equivalence relation, possibility of equivalence between the two objects cannot be accurately proposed by using a limited torrent relation. In a probabilistic rough set model, set-valued decision reduction method and dominance rough set-valued order information systems are discussed. The probability dominance rough set model mainly deals with dominance relation analysis and probability dominance relation, and an example for this model is also discussed here. In that example, probability dominance classes are found [2].

In this research work, the authors have discussed the evaluation of a model supported set-based data estimation system. It describes the disease processes on a detailed level. The biomedical information and the measurements have different quantities and qualities. New methods are required for this process. So, an approach that integrates the signals for short- and long-term processes has been decided for different frameworks. In the end, set-based estimation methods are combined for pathways and that framework is demonstrated by Jak-STAT3 and MAPK trans-signaling. It can be considered as a dynamic model. In the initial experimental setup, artificial knowledge is generated from linguistic and it is logically descriptive, and the generated logical descriptive linguistic is used to measure the model capability in order to uncover fundamental relative reasoning, algorithmic structures, and quantification.

A framework that requires one dynamic model for the short time scale processes and patient data for short- and long-term scale processes has been proposed, and it can be considered as a classification method. It is continued with mathematical formulation, parameters, and dynamic processes. Two algorithms are presented and they describe how the method takes the correct stratification and how it is applied. The first step is to find the inputs for the classification methods. In the second step, they predict the long-term process from the parameters that have been determined by the first step. This approach combines the set-based estimation framework. The first algorithm is named as set-based classifier training. The inputs of this algorithm are short- and long-term scale data of patients and the model has been described by a dynamic untrained classifier. The output is a classifier that would be trained. This algorithm performs the set-based estimation of parameters using a dynamic model. It stratifies the patients into risk subcategories. It splits the patients into a group for validation and test. This algorithm trains the chosen classifier and verifies the classifier's quality. The second algorithm is patient stratification. The inputs of that algorithm are the same as the inputs of the first algorithm. The output is a patient-specific risk category prediction. This algorithm performs the set-based estimation of parameters using a dynamic model. It takes set-based estimation results as input to the trained classifier and verified with the stratification results. Jak-STAT3 pathway model initiates trans-signaling. The MAPK pathway model plays an important role in differentiating the cells. Finally, a modeling framework that could be considered as an extension of methods has been provided. The disease processes are transformed by a classification algorithm to provide a prediction. It has been aimed to provide feasibility sets to describe as a parent-class and explain the uncertainty of data [3].

The authors in [4] have proposed the enquiry system and by using which the customers can ask questions about a particular service and obtain relevant information about the system. To implement an enquiry system, a question and answer technology is needed. For this, a machine learning algorithm has been suggested and by which a list of training data consisting of a certain pair of assumed questions and relevant answers to those questions can be prepared. The first step is to put together a solution set for a carrier and the solution set must comprise all the feasible solutions approximately. By the usage of consumer behavior gadget version and the service-feature model, they can correctly gather an understanding answer set on the provider. According to this enquiry device, the question and solution technology plays a vital role. The query and solution technology is made to have many APIs. For the usage of machine studying strategies or algorithms, we need to put together the schooling facts that include all viable pairs of solution units in addition to question units. The first-rate and quantity of question and answer set have been confirmed successfully as well as it guarantees accuracy and suitability. The buying stages are brought by organizing expertise records that are required for an enquiry machine.

Buying stages technique is primarily based on the patron conduct version and the expertise facts are constructed for each and every buying stage. By using this model, a service characteristic model has been designed, wherein steps are proposed for gathering provider factors. While training, the carrier factors are fed to derive the required knowledge. Then, the knowledge data are prepared for a solution set. It is shown that their schooling facts satisfy accuracy. They should accumulate the service elements exhaustively, and the domain experts can gather knowledge data with the aid of their personal way. By making use of the proposed gadget in this mission of Q&A-based totally inquiry system, especially for the usage of client conduct and service feature modeling, it is shown that they are able to gather nearly all the answers that the clients may ask and they avert the case that offers an end result as an unsuitable answer for the question that the customers ask. Another factor is that the developer and service company will outline a brand new solution set for the query that the gadget cannot respond to correctly. In many different systems, the provider can collect all the questions on the middle of the facts. It has been concluded that they may adjust their proposed method to arrange information facts for an enquiry gadget with the aid of FAQs [4].

Grammatical error correction (GEC) is the process of sleuthing and correcting grammatical errors in the text written by non-native English writers. Unlike construction of machine learning classifiers for specific error sorts (e.g., determiner or preposition errors), "translating" an incorrect grammar sentence into a correct one has been considered to handle all error sorts at the same time. Applied math MT (SMT) has been effectively used for GEC, as in contestable by the top-performing systems inside the CoNLL-2014 shared task.

The work by the authors in [10] tends to use the RNN search model that contains a bi-directional RNN as an associate in nursing encoder and an attention-based decoder. Forward RNN and backward RNN are combined, and it is known as the bi-directional RNN encoder. The forward RNN reads the provided sentence from the main word to the last, and the backward RNN reads the provided sentence in reverse order. The historical and future data are captured with the bi-directional RNN encoder. The attention-based model allows the decoder to familiarize the foremost relevant data within the sentence provided, instead of fundamental cognitive process.

11.3 Set-based Model for Reinforcement Learning Design for Medical Data

In this approach, for non-linear unknown dynamic systems, an approximate solution for optimal control problems has been proposed. The main technique used is the strategy of reinforcement learning, which combines the extremer phasor seeking controllers. This approach of learning is mainly used to find out the value function for the control problem of optimal interest. Here, the approximate optimal controller is implemented using a seeking controller of extreme phasor. The optimal control problems with reasonable approximations are provided without any need of parameterization for a control system of non-linearity. Many model-based control systems play important roles in the recent developments of control systems based on reinforcement learning, which is mainly used in machine learning algorithms for providing various solutions for problems and also to provide decision support systems for solving the problem.

In reinforcement learning, the main technique based on the model used is set-based least squares of Q-learning. By this technique, the unknown value function is determined or identified by using the actor–critic Q learning methodologies, and it also provides the feedback of a corresponding optimal state. Using this technique, the Hamilton–Jacobi bell constraint is created with regard to the unknown value function. This approach is mainly referred to as reinforcement learning of the actor–critic approach. This actor–critic approach usually results in the difficulty of the estimation of adaptive parameter because it usually requires very conservative and highly complex conditions of the corresponding excitations. This approach is extremely challenging. So, the exploitation of such parameterization techniques is done by the set-based model. The main use of a set-based model is identifying the errors in terms of ec(x) and ea(x), which are associated with the unknown value function with regard to the parameterization. Hence, the approximation errors referred to this approach are made negligible by the set-based model, which results in the analysis of more complications.

In this approach, the role-based stabilization is done by the set-based model of the least squares learning. This learning involves set-based estimation of the phasor with stabilization. So, first, the closed-loop system stabilization is considered with the feedback of the state. If it is greater than zero, it produces the positive-definite constant and the positive-definite matrix, which is used for the stabilization of the originality of the closed-loop system based on the non-linearity of the system and it is stabilized globally asymptotically. This approach results in an efficient learning algorithm that is used for the trajectories of the learning algorithm. Hence, this technique results in the effectiveness of non-linearity of the optimal control problem. Set-based model is the recent approach that has started from 2005. It is one of the alternative set theoretic models. The model provides the analysis of context terms for medical data with the incorporation of vectorial analysis [5]. The idea is to follow the process of mutual dependency in order to improve the search results. The text that is present may signify different forms of levels for the terms that are used in the medical context. It may be of level A, level B, level C, and level D accordingly. The main context is to find the occurrence and significance of those repeated terms with different contextual patterns.

Initially, this IR model uses the advantage of term dependence with collections. The set-based model contains three important terms that are termsets, ranking computation, and closed termsets. The formula of the set-based model contains closed termset, the inverted frequency in the collection, and the number of documents in the collection. The closed termsets are the frequent termsets that are the largest termsets. The maximal termsets are also the frequent termsets, but they are not subsets of any other frequent termsets. Termset is the subset of terms in the termset collection. The collection of sets is called vocabulary set.

The first step is to give the notations to each word. Then, the letters refer to the index terms. Next, the query is considered. The query generates the termsets that only need to be considered at query processing time. The termset comprises n terms, and then it is called an n-termset. Frequent termsets are used to minimize the number of termsets in order to consider longer queries. Computation of termsets is done in the set-based model, but the ranking computation is based on the vector model. It considers a query and a document. The document norm finds hard to compute in termsets space present in the set-based model [6].

The simplification of ranking computation is allowed by frequent termsets. Smaller numbers of termsets are used to solve this problem for ranking

Table 11.1   Document terms representation.

d intersect q	Documents	Rank
3	$d1, d2$	1
2	$d3$	3
1	$d4$	4

computation. The closure of termset is the largest termset, which is called closed termset. It encapsulates the smaller termsets that occur in the repeated set of documents. If there is restriction of computation given to the closed termsets, then the reduction of query time will be more beneficial. The usage of smaller number of termsets leads to sharp reduction processing time. The ranking of document is based on the query terms.

11.3.1 Case study

Consider the set of index terms for the query q and document d, respectively. Ranking is based on the cardinality of d intersect q, i.e., with regard to the number of common terms to the document and query.

$$R(d, q) = | d \ intersect \ q |,$$

where $q = \{$ diagnosis, blood levels, scan reports$\}$; so Table 11.1 is the output.

By using this document set, compute the rank for the given query. Consider the query as "To do be it."

In order to simplify the notation, we can define the following:

$k_a = \text{to}$
$k_b = \text{do}$
$k_c = \text{is}$
$k_d = \text{be}$
$k_e = \text{or}$
$k_f = \text{not}$
$k_g = \text{I}$
$k_h = \text{am}$
$k_i = \text{what}$
$k_j = \text{think}$
$k_k = \text{therefore}$
$k_l = \text{what}$
$k_m = \text{let}$
$k_n = \text{it}$

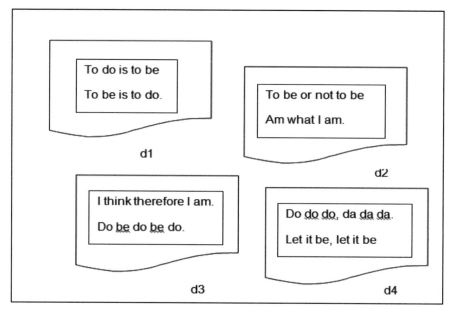

Figure 11.1   Document representation (referred by Chapter 3: Modeling, Baeza-Yates & Ribeiro-Neto, Modern Information Retrieval, 2nd Edition – p. 94).

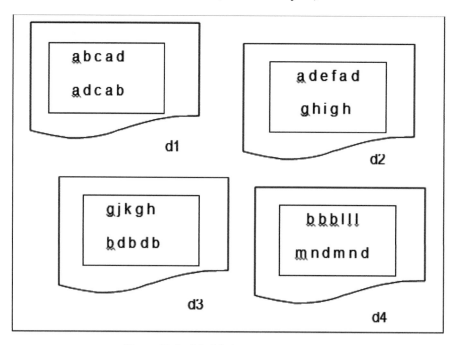

Figure 11.2   Modified term representation.

Table 11.2   Termsets and their corresponding document classification.

Let q = "to do be it," i.e., $q = \{a, b, d, n\}$

For this query, the vocabulary set is as follows:

a = medical term 1; b = medical term 2; d = medical term 3;

n = corpus.

Termset	Set of terms	Documents
Sa	$\{a\}$	$(d1, d2\}$
Sb	$\{b\}$	$\{d1, d3, d4\}$
Sd	$\{d\}$	$\{d1, d2, d3, d4\}$
Sn	$\{n\}$	$\{d4\}$
Sab	$\{a, b\}$	$\{d1\}$
Sad	$\{a, d\}$	$\{d1, d2\}$
Sbd	$(b, d\}$	$\{d1, d3, d4\}$
Sbn	$\{b, n\}$	$\{d4\}$
$Sabd$	$\{a, b, d\}$	$\{d1\}$
$Sbdn$	$\{b, d, n\}$	$\{d4\}$

The document representation and modified document representation is shown in the Figure 11.1 and 11.2 respectively. Further, let the letters a, ..., n refer to the index terms k_a, ..., k_n.

Table 11.2 represents the termsets and their corresponding document classification representation.

Given with maximum of 15 termsets, 11 termsets extraacted can be formed with the terms in q.

Note that in our collection, out of the maximum of 15 termsets, that can be formed with the terms in q.

Assuming 2 as the threshold for the frequency of termsets, we can calculate frequent termsets for all based on the query $q = \{a, b, d, n\}$ with the following steps:

1.   Calculate the significant 1-terms and its corresponding list (inverted) data:

$Sc = \{d1, d2\}$
$Sd = \{d1, d3, d4\}$
$Se = \{d1, d2, d3, d4\}$

2.   Combine the IV (inverted list data)

3.   Compute frequent 2-termsets:

$Scd = \{d1, d2\}$
$Sde = \{d1, d3, d4\}$

4.   As there are no frequent three termsets found, stop

5.   Since there exists only five frequent items in the data collection

6.   The IV for the term frequent n-termsets can be evaluated along with the interval leveled data, and, therefore, one index is required as well as it is standard IV for any such IR system.

7.   For 4–5 terms of short queries, this is reasonably fast.

11.3.2 Rank computation

The ranking system is similar to the vector space model but fixes the termsets rather than index sets. If the given query is taken as q, and assume that $\{S1, S2, \ldots\}$ is the set of all termsets that originated from q. Ni is the number of documents that contain the occurrence of the termset Si. N is the count total for the documents found. Let Fi, j be the frequency of termset Si in document dj.

Queries q and a and document dj are represented by using the vectors in a $2t$-dimensional space:

$$\sim dj = (W1, j, W2, j, \ldots, W2t, j)$$

$$\sim q = (W1, q, W2, q, \ldots, W2t, q).$$

Query q will have the rank corresponding to the document dj and is computed as follows:

$$\text{sim}(dj, q) = (\sim dj \sim q)/|\sim dj| \times |\sim q| = (\text{summation}) \; Si \; Wi, j \times Wi, q/|\sim dj| \times |\sim q|.$$

For termsets, if not in the query q, $Wi, q = 0$. The document norm $|\sim dj|$ is hard to compute in the space of termsets. Thus, its computation is restricted to 1-termsets. Let q be the set containing $\{a, b, d, n\}$ and $d1$. The norm document with respect to 1-termsets will be

$$|\sim d1| = \text{square root } (W2a, 1 + W2b, 1 + W2c, 1 + W2d, 1)$$
$$= \text{square root } (4.752 + 2.442 + 4.642 + 2.002 = 7.35$$

$$\text{Rank of } d1 \text{ will be } \text{sim}(d1, q) = 5.71.$$

The Weight computation and its termset representation is shown in Table 11.3. Hence, if the restriction is made with regard to ranking computation, then there would be a complete reduction in the query analysis time [8].

Table 11.3   Weight computation and its termset representation.

if $F_{i, j} = 0$

$W_{i, q}$ value is computed for the pair $[S_i, q]$
Consider query $q = \{a, b, d, n\}$
document $d1 = $ "$a\ b\ c\ a\ d\ a\ d\ c\ a\ b$"

Termset		Weight (by using above formula)
Sa	Wa, 1	4.75
Sb	Wb, 1	2.44
Sd	Wd, 1	2.00
Sn	Wn, 1	0.00
Sab	Wab, 1	4.64
Sad	Wad, 1	3.17
Sbd	Wbd, 1	2.44
Sbn	Wbn, 1	0.00
Sdn	Wdn, 1	0.00
Sabd	Wabd, 1	4.64
Sbdn	Wbdn, 1	0.00

Hence, the validation can be made in accordance with the text analytics flow model or the stepwise approach to text data classification. The rankings may vary in accordance with the data chosen for evaluation in medical context. The sample text embedding for the selected textual data is shown in Figure 11.3

11.3.3 Tools for evaluation

Data analysis plays a significant role in different types of applications especially for medical data analysis. The extraction of textual data and its variants need to be carefully analyzed in different slots and in its stream processing modes. Analyzing the set-based models and deriving its impact are very essential [7]. The following are the different forms of tools that aid in processing the textual data for processing the real-time information and deriving set-based theoretic models:

- SAS
- Rapid miner
- Voycant tools
- Watson analyzer
- Microsoft's Cognitive Services suite
- QDA Miner's WordStat

Figure 11.3   Text embedding for the selected textual data.

11.3.4 Applications

The variants of text-based models are significantly used in different sectors for conceptual analysis of text design and extraction. Significant analysis has to be made in making the data to be exact and confirming a decision at the best level [9]. Applications include bio-informatics, signal processing, content-based retrieval, and speech recognition platforms.

1.   Bio-informatics: It is a tool for understanding biological data that combine biology, computer science, information engineering, mathematics, and statistics. This provides a complete set of model extraction in medical data analysis and it is of type semi-structured in an unstructured context.

2.   Bio-signal processing: It is a signal in living beings and that can be continually measured and monitored.

3. Content-based image retrieval: It is the problem of searching for digital images in large databases.

4. Facial recognition system: It is the system capable of identifying or verifying a person from a digital image or a video frame from a video source.

5. Speech recognition system: It is the system that enables recognition and translation of spoken language into text by computers.

6. Technical chart analysis: It is done forecasting the direction of prices through the study of past market data, primarily price and volume.

11.4 Conclusion

Use cases of text data analysis are deployed in different fields of practice. Managing the integrity and the significance of the data are the main concerns. Determining the exact occurrence and the reflection of textual data seems to be an important paradigm in text data classification. This chapter provides a pathway for evaluating the set-based model in the information retrieval modeling process. Here, a sample text data in the medical domain have been evaluated for processing the set-based IR model with its processing steps and procedures. In addition to other IR models, set-based models play a significant role in determining the ranking of documents with 95% accuracy level for all types of text processing engines. Different sorts of tools may find variance in evaluating the ranking of documents for a given collection, but the exact realm lies at the term frequency and the inverse document frequency having the exact deviation for the data that are used. In future, these models may impact some changes on the search engines with regard to the ranking of text data by adhering the parametrics and principle of text metrics for analyzing medical data.

References

[1] Rudolph, N., Streif, S., &Findeisen, R. (2016). Set-based Experiment Design for Model Discrimination Using Bilevel Optimization. IFAC-PapersOnLine, 49(26), 295–299. doi:10.1016/j.ifacol.2016.12.142

[2] SuoZhongying, Cheng Siyi, &RenJinshen. (2016). Probability rough set model based on the semantic in set-valued information system. 2016 2nd IEEE International Conference on Computer and Communications (ICCC). doi:10.1109/compcomm.2016.7924903

[3] Rudolph, N., Andonov, P., Huber, H. J., &Findeisen, R. (2018). Model-supported Patient Stratification Using Set-based Estimation Methods. IFAC-PapersOnLine, 51(18), 892–897. doi:10.1016/j.ifacol.2018.09.233

[4] Takeuchi, H., Masuda, S., Miyamoto, K., &Akihara, S. (2018). Obtaining Exhaustive Answer Set for Q&A-based Inquiry System using Customer Behavior and Service Function Modeling. Procedia Computer Science, 126, 986–995. doi:10.1016/j.procs.2018.08.033

[5] Choudhary, U., Dhurandher, S. K., Kumar, V., Woungang, I., & Rodrigues, J. J. P. C. (2018). Online Signature Verification Using the Information Set Based Models. 2018 IEEE 32nd International Conference on Advanced Information Networking and Applications (AINA). doi:10.1109/aina.2018.00158

[6] Doszkoos, T. E. (n.d.). From research to application: The cite natural language information retrieval system. Lecture Notes in Computer Science, 251–262. doi:10.1007/bfb0036350

[7] Pôssas, B., Ziviani, N., Meira, W., &Ribeiro-Neto, B. (2002). Set-based model. Proceedings of the 25th Annual International ACM SIGIR Conference on Research and Development in Information Retrieval - SIGIR '02. doi:10.1145/564376.564417

[8] Sheik Abdullah A., Akash K., Bhubesh K. R. A., &Selvakumar S. (2021). Development of a Predictive Model for Textual Data Using Support Vector Machine Based on Diverse Kernel Functions Upon Sentiment Score Analysis. International Journal of Natural Computing Research, 10(2), 1–20. doi:10.4018/ijncr.2021040101

[9] Abdullah, A. S., Parkavi, R., Karthikeyan, P., &Selvakumar, S. (2021). A Text Analytics-based E-Healthcare Decision Support Model Using Machine Learning Techniques. Smart Computational Intelligence in Biomedical and Health Informatics, 169–182. doi:10.1201/9781003109327-12

[10] Feng Tao, Millard, D., Woukeu, A., & Davis, H. (2005). Managing the semantic aspects of learning using the knowledge life cycle. Fifth IEEE International Conference on Advanced Learning Technologies (ICALT'05). doi:10.1109/icalt.2005.195

[11] Salton, G., Wong, A., & Yang, C. S. (1975). A vector space model for automatic indexing. Communications of the ACM, 18(11), 613-620.

Index

A

AR 110–112, 116–118, 120–121, 123–124, 126
ARIMA 110–112, 115–118, 121–122, 124, 126–127
association rules 188, 189

C

Cancer 3, 16, 55, 65, 70, 91–97, 105, 110, 149–151, 153, 159, 162–163, 165
chest X-ray (CXR) images 54, 67, 69, 73, 75–78, 81–85
classification 3, 5–8, 12–14, 16, 22–24, 27, 29, 31, 39, 52–53, 55–58, 60, 62–65, 67, 69–70, 73–79, 81–85, 91, 94–95, 97–98, 101, 104–105, 131–132, 136–137, 140, 149–154, 156–159, 163, 165, 172–173, 180, 187, 190–191, 197, 199, 201
class imbalance 16
CNN 39, 48, 53–54, 56–57, 61–62, 65–69, 73–76, 78–79, 81–82, 84–86, 94, 137, 139, 153, 155
Covid-19 1, 16, 35, 37, 48, 54–57, 73–76, 82–83, 85–86, 109–112, 117–118, 123, 127
CS 150, 153, 155, 159–160, 165

D

data mining 20, 23, 27, 29, 136, 189
dataset 1–17, 19, 22, 24, 27, 29–32, 54–57, 59, 63–64, 66–68, 70, 75–76, 80, 82, 83, 85, 95, 101–102, 104, 110, 113, 127, 134, 139–140, 150–151, 153–155, 157–159, 162, 170, 172, 175, 178–180, 182–183
deep learning 51–52, 54–57, 70, 74–76, 78–79, 93, 132–133, 139
dimensionality reduction 20, 57–58, 63, 136, 150–151, 153–154, 159, 169–171, 178–179, 182–183
document set 195

E

ECG 131–132, 136–137, 139–140
EEG 131, 132, 139–140
Electronic health records 131–132, 140

F

feature selection 19, 22–24, 62, 75, 106, 137, 150–153, 159–160, 162–163, 165, 169–178, 182

I

index term 188–189

L

LR 110, 118–119, 121, 124
LSTM 91–92, 95, 101–103, 105

M

MA 110, 115–116, 118, 120–121, 124, 126–127
Machine learning 1–3, 6–8, 36, 51–52, 54, 56–58, 61–62, 65–70, 91–93, 101, 104–105, 131–136, 138–141, 169–172, 176–180
Medical 1–3, 8, 16, 20, 35–36, 39–40, 42, 45, 48, 51–54, 56–59, 61–62, 64–65, 68–70, 73–74, 76, 85–86, 109, 131–133, 140, 151, 169, 187–188, 193–194, 197, 199–201
medical data 3, 20, 64, 70, 131, 140, 169, 187–188, 193–194, 199–201
medical imaging 51, 73–74, 132
microarray data 149–153, 155, 158–159, 163
modified dynamic adaptive particle swarm optimization (MDAPSO) 19, 26
monitoring 20, 35–37, 41, 46, 138, 140

N

naive Bayes 6, 7, 23, 91, 101, 105, 136

O

oversampling 2, 4–6, 10, 12–17

P

Particle swarm optimization (PSO) 19, 22
PPG 131–132, 137–140

PR 110

pre-processing 2, 5, 8, 53, 56–57, 63, 66–68, 70, 95, 97, 132, 140–141, 157, 172

Q

query 187–189, 192, 194–195, 197–199

R

random forest 8, 11–16, 36, 42–43, 47–48, 59–60, 64, 70, 91, 101–103, 105, 158, 180
ranking 160, 175–177, 188, 194–195, 198, 201
RFE 150, 153, 155, 159–160, 162–163, 164, 165, 166
RMSE 110, 116, 118, 121, 124, 129

S

SARIMA 110–112, 116–118, 121–122, 124, 127
SMOTE 2, 6, 10, 13–17
SVM 23, 24, 29–31, 56, 59–63, 75–76, 82–86, 94–95, 110–112, 114, 117–119, 121–122, 124–125, 127, 136–137, 150, 152–153, 155–156, 159, 162–165

T

Term weight 188

U

undersampling 2, 4–6, 10, 12–17

V

vital signs 36, 41–42, 46

X

XGBoost 91, 95, 101–103, 105

Author Biographies

E. Ramanujam (Senior Member, IEEE) is currently working as an Assistant Professor with the Department of Computer Science and Engineering, National Institute of Technology Silchar, Assam, India. He has more than 10 years of teaching and research experience. He received the Ph.D. degree in information and communication engineering from Anna University, Chennai, India, in the year 2021. He has published research papers in peer-reviewed international journals from IEEE, Inderscience, and IGI Global publisher and national and international conferences organized by IEEE. He has authored several chapters in the refereed edited books of IGI and Springer. He also acts a Reviewer for international journals such as IEEE, Springer, Elsevier, Emerald, and IGI Global Publisher.

T. Chandrakumar is currently an Associate Professor with the Department of Applied Mathematics and Computational Science, Thiagarajar College of Engineering, Madurai, Tamil Nadu, India. He received the Ph.D. degree from Anna University specializing in computer science. His research interests include software engineering, analytics, enterprise resource planning (ERP), and software quality. He has completed a few projects receiving funding from University Grants Commission (UGC) India. He has been regularly funded by DST (Department of Science and Technology) and DRDO (Defence Research Development Organization) for organizing national-level workshops and seminars. He has published more than 15 research papers in refereed conferences and journals such as *Computer Standards & Interfaces* (Elsevier-SCI), *International Journal of Project Management* (Elsevier-SCI), *International Journal of Enterprise Information Systems* (IGI-USA), *International Journal of*

205

Informatics and Communication Technology, and *International Journal of Business Information Systems*. He has authored 05 book chapters in the refereed edited book of Springer from January 2014 till date. He also serves as a Reviewer for the *International Journal of Project Management* (Elsevier), *Computer Standards and Interfaces* (Elsevier), *International Journal of Industrial Engineering: Theory, Applications, and Practice*, etc. He has guided many postgraduate scholars and teaches several courses on computer applications and data science engineering.

D. Sakthipriya is currently doing full-time Ph.D. research at Anna University, Chennai, India. She has published Indian Patent and research articles in international journals, book chapters, and international conferences. Her research area includes machine learning and deep learning in agriculture domain. She also have more than 5 years of teaching experience in computer science-related subjects for undergraduation and post-graduation.

J. Shanthalakshmi Revathy received the B.E. degree in information technology from Mohamed Sathak Engineering College and received the M.E. degree in computer science and engineering from the PSG College of Technology. Her areas of research interest include data science and evolutionary computing. She is currently working as an Assistant Professor with the Velammal College of Engineering and Technology, Madurai with a teaching experience of 10 years.

J. V. Anchitaalagammai received the Ph.D. degree in computer science, with specialization in information and communication engineering, from Anna University, Chennai, India, in 2017. She has acquired enormous working experiences as a Lecturer and a Researcher in computer science. She is currently an Associate Professor with the Department of Computer Science and Engineering, Velammal College of Engineering and Technology, Madurai, Tamil Nadu, India. She has published several research papers in prestigious international journals and conference proceedings. She has been serving as the Reviewer for several journals. Her research interests span a broad range of interesting topics, including wireless networks and network security, blockchain, big data analytics, and the Internet of Things.

S. Hariharasitaraman completed his doctorate on the topic of analyzing data integrity schemes in cloud computing from KARE. His doctorate dissertation outlines various research directions and opened up future avenues in designing secure data integrity protocols. He received the M.Eng. degree in computer science and engineering from Anna University in the year 2005, with master's dissertation outlines in designing and analyzing software infrastructures for the internetworking environment. He received the B.Eng. degree in the field of computer science and engineering in the year 2003. He has more than 28 publications, 1 book chapter, and 2 books in national and international conferences, and peer-reviewed international journal proceedings indexed in Scopus, SCIE, and SCI databases. He is the recipient of Elsevier Publons Certified Journal Reviewer award and reviewer for the *Journal of Supercomputing* (Springer). He is an active member of various technical national and international bodies like the Information Security Awareness Council, MeitY, GoI, ACM, IEEE, ISTE, CSI, and IAEng. His areas of research interest include computing, security, ambient, and artificial intelligence.

Premanand Ghadekar (http://orcid.org/0000-0003-3134137X) received the Ph.D. degree from SGBA University, Amravati, India. He received the M.Tech. degree in electronics (computer) from the College of Engineering, Pune, India, in the year 2008. He received the B.E. degree in electronics and telecommunication engineering from the Government College of Engineering, Amravati, India, in 2001. In 2003, he joined the Department of Computer Engineering, Vishwakarma Institute of Technology, Pune, Maharashtra, India. He is currently working as a Professor and Head of Information Technology Department, VIT Pune. His areas of research are IoT, image processing, computer vision, video processing, dynamic texture synthesis, machine learning, deep learning, and embedded system. He has contributed 12 papers in international conferences, 15 papers in international journals, and 8 papers in Springer book series. He is a life member of ISTE, a member of CSI, and a member of IEEE. He has completed a research project of 2 lakhs.

Pradnya Katariya completed graduation in information technology from the Vishwakarma Institute of Technology, Pune, India, in the year 2021. She is currently working with Credit Suisse as a Technology Analyst. She has been working in technologies like Java, Spring Boot, Jenkins, React, and Angular.

Shashank Prasad completed graduation in information technology from the Vishwakarma Institute of Technology, Pune, India, in the year 2021. He is currently working with Deutsche Bank as a Technology Analyst. He has been working in technologies like Cloud, Java, Spring Boot, Angular, and React.

Aishwarya Chandak received the bachelor's degree in information technology from the Vishwakarma Institute of Technology, Pune, India, in 2021. She is currently working as a Technology Analyst with Deutsche Bank.

Aayush Agarwal completed graduation in information technology from the Vishwakarma Institute of Technology, Pune, India, in the year 2021. He is currently working with NICE Actimize as an Associate Software Engineer. He has been working in technologies like Java, SQL, Spring, and Spring Boot.

Anupama Choughule completed graduation in information technology from the Vishwakarma Institute of Technology, Pune, India, in the year 2021. She is currently working with Pragmasys Solutions as a Software Developer Engineer. She has been working in technologies like, CRM, DOTNET, C#, Java, and SQL.

V.Thamilarasi received the Ph.D. degree in computer science, with specialization of Digital Image Processing from Periyar University, Salem, Tamilnadu, India, in 2022. She has more than 7 years of working experience as Lecturer, Assistant Professor. She is currently an Assistant Professor in the Department of Computer Science, Sri Sarada College for Women(Autonomous), Salem, Tamilnadu, India. She has published several research papers in prestigious international journals, 2 book chapters and one of her paper received best paper award in 2021. She has been serving as the Reviewer for Elsevier, Scopus journals. Her area of research interest include Medical Image Processing, Machine Learning, Computer Vision and Deep Learning. (email id: tamilomsiva@gmail.com)

Dr. R. Roselin currently working as an Associate Professor of Computer Science since 2001. Her area of research includes Digital Image Processing, Data Analytics and Signal Processing. She published many papers in International Journals, and edited chapters in books and attended many conference and completed MOOC based Faculty Development Programmes. (email id: roselinjothi@gmail.com). Her google scholar link is: https://scholar.google.com/citations?user=iVkHyC0AAAAJ&hl=en&authuser=1

Nam Anh Dao received the B.S. degree in applied mathematics and the Ph.D. degree in physics-mathematics from the University of Moldova, in 1987 and 1992, respectively. He was involved in various international software projects. He is currently teaching at Electric Power University.

His research interests include intellectual intelligence, image processing and pattern recognition, machine vision, and data science. His main works cover pattern recognition and image analysis, medical imaging, and machine learning with emphasis on computer vision. He also served or is currently serving as a Reviewer for many important journals and conferences in image processing and pattern recognition.

Anh Ngoc Le is currently a Director of Swinburne Innovation Space, Swinburne University of Technology, Vietnam. He received the B.S. degree in mathematics and informatics from Vinh University and VNU University of Science, respectively. He received the master's degree in information technology from the Hanoi University of Technology, Vietnam. He received the Ph.D. degree in communication and information engineering from the School of Electrical Engineering and Computer Science, Kyungpook National University, South Korea, in 2009. His general research interests include embedded and intelligence systems, communication networks, the Internet of Things, image/video processing, AI, and big data analysis. On these topics, he has published more than 40 papers in international journals and conference proceedings. He served as a keynote, TPC member, session chair, and reviewer of international conferences and journals (email: *ngocla2@fe.edu.vn*).

Le Manh Hung received a Bachelor's degree from Hong Duc University and Master Degree in IT from Posts and Telecomms Inst of Technology (PTIT), Ha Noi, Viet Nam. He is now a lecturer in Faculty of Information Technology, Electric Power University, Hanoi, Viet Nam. His current research interests are in medical data processing and analysis, Pattern Recognition, NLP, Expert Systems, Machine Learning, Block-Chain, and Big Data. In the field of bioinformatics and computational biology precision medicine, genomics, gait recognition, and public health.

P. Shanmuga Sundari received the bachelor's degree and the master of computer application degree in computer science from Theivanai Ammal College for Women in 2000 and 2003, respectively, the master's degree in computer engineering from the Jayaram Institute of Engineering and Technology in 2008, and the philosophy of doctorate degree in computer engineering from the Vellore Institute of Technology in 2021. She is currently working as an Associate Professor with the Department of Computer Science and Engineering, Sri Venkateswara College of Engineering and Technology. Her research areas include machine learning, deep learning, social network analysis, and machine learning approach for mechanical application.

J. Jabanjalin Hilda received the bachelor's degree in instrumentation and control engineering from Madurai University, the master's degree in computer engineering from the Vellore Institute of Technology, and is currently working toward the philosophy of doctorate degree in computer engineering from the Vellore Institute of Technology. She is currently working as a Senior Assistant Professor with the School of Computer Science and Engineering, Vellore Institute of Technology. Her research areas include cloud computing, machine learning, and data mining.

S. Arunsaco received the bachelor's degree in mechanical engineering from Anna University, the master's degree in M.Tech CAD/CAM from the Vellore Institute of Technology, Vellore, India, and the philosophy of doctorate degree in mechanical engineering from the Vellore Institute of Technology, Vellore, India. He is currently working as an Associate Professor with the Department of Mechanical Engineering, Sri Venkateswara College of Engineering and Technology (Autonomous). His research areas include fuel cells, computational fluid dynamics, and the implementation of machine learning approach to mechanical application.

Pradumn Kumar Mishra is currently working toward the bachelor's degree in computer science and communication engineering with the Kalinga Institute of Industrial Technology. He is currently working as a Machine Learning and Robot Operating Systems Developer with SwayamChalit Gaadi, the only autonomous vehicle racing team in the entire Odisha. His research areas include data science, deep learning, machine learning, robotics, and automation using ROS. He has been constantly working on self-driving technology and is currently leading the software team.

Abhisek Omkar Prasad is currently working toward the bachelor's degree in computer science and communication engineering with the Kalinga Institute of Industrial Technology. He is currently working as a Machine Learning, Reinforcement Learning, and Robot Operating Systems Developer with SwayamChalit Gaadi, the only autonomous vehicle racing team in the entire Odisha. His research areas include data science, deep learning, machine learning, robotics, edge computation, and automation using ROS. He has been constantly working on self-driving technology and currently leading the team.

Dibyani Banerjee is currently working toward the bachelor's degree in computer science engineering with the Kalinga Institute of Industrial Technology. She is closely working with the placement cell of KIIT DU as a Research Associate. She is currently interning with an MNC known by the name Akon International as an Automation Developer, working on autonomous vehicles. Her areas of research interests are in algorithms, machine learning, artificial intelligence, robotics, computer vision, and related technologies. In addition to this, she is an active member of AIESEC in Bhubaneswar, demonstrating integrity and activating leadership skills. She has also been volunteering at The National Service Scheme (NSS) for the betterment of society.

Megha Singh is currently working toward the bachelor's degree in computer science and system engineering with the Kalinga Institute of Industrial Technology. Her areas of research interests are in machine learning, artificial Intelligence, natural language processing, data science, and algorithms. She is an active member of YRC KIIT and also makes contributions to the National Service Scheme (NSS) for the betterment of society.

Shubhi Srivastava is currently working toward the bachelor's degree in computer science and system engineering with the Kalinga Institute of Industrial Technology. Her areas of research interests are in machine learning, artificial intelligence, natural language processing, data science, electric vehicle, and web development. She is an active member of KIIT Hack Club. She enjoys reading about new technologies as well as using them in real-life scenarios.

Abhaya Kumar Sahoo is currently working as an Assistant Professor with the School of Computer Engineering, KIIT Deemed to be University, Odisha, India. He received the Ph.D. degree from KIIT University. He has published more than 15 papers in reputed international journals and conferences. His areas of research interests include recommender system, data analytics, image processing, and parallel computing. He is the life member of CSI and IET.

Sonkoju Nagarjuna Chary received the bachelor's degree in electronics and instrumentation engineering from Jawaharlal Nehru Technological University, Hyderabad, India, in 2008, and the master's degree in process control & instrumentation from Annamalai University, India, in 2011. He is currently working as an Assistant Professor with the Department of Electronics & Instrumentation Engineering, VNR Vignana Jyothi Institute of Engineering & Technology,

Hyderabad, India. His research areas include biomedical engineering, deep learning, and EHR analysis.

N. Vinoth received the B.E. degree in the year 2002 in electronics and instrumentation engineering and the M.Eng. Degree in process control and instrumentation in 2007 from Annamalai University, India. He received the Ph.D. degree in 2015 from Annamalai University in the field of instrumentation. He has 17 years of teaching experience. He is currently working as an Associate Professor with the Department of Instrumentation Engineering, M.I.T Campus, Anna University Chromepet, Chennai, India. His current research interests include biomedical engineering, machine learning and drives, and deep learning.

Kiran Chakravarthula received the B.Tech. degree in electronics and instrumentation engineering from Acharya Nagarjuna University, India (Bapatla Engineering College). He pursued his higher education at Louisiana Tech University, USA, where he obtained three Master of Science degrees – in engineering (biomedical engineering), microsystems engineering, and applied physics – followed by a Ph.D. degree in engineering (engineering physics track). He completed his PG Diploma in innovation and design thinking from EMERITUS Institute of Management, Singapore, in collaboration with MIT Sloan School, Tuck School of Business, and Columbia Business School. He currently serves as an Associate Professor with VNR Vignana Jyothi Institute of Engineering and Technology, Hyderabad, India, where he also heads the Global Relationships Office, handles the external innovation collaborations, and convenes the Institute Innovation Cell and is also the Institute's Press & Media Coordinator. His research in biomedical engineering and high energy physics has been published in reputed international journals. Dr. Chakravarthula has been invited to deliver talks on diverse topics including lasers, ethics, design thinking, internet of things and ethics, ethics in healthcare applications, professional ethics, and stress management at various national-level events.

Amrutanshu Panigrahi is currently working as a Research Scholar with the Department of CSE, Siksha O Anusandhan (Deemed to be University), Bhubaneswar, Odisha, India. He received the master's degree in information technology from the College of Engineering and Technology, Government of Odisha and is currently working toward the Ph.D. degree with the Department of CSE, SOA University, Bhubaneswar, India. Mr. Panigrahi has published 14 research articles in peer-reviewed and Scopus indexed journals, 6 international conferences, and 3 Scopus indexed book chapters. He has been working as a Reviewer for multiple international journals. He has more than seven years of academic experience and his current research interests include blockchain technology, cloud computing, machine learning, and deep learning.

Manoranjan Dash is currently working as an Associate Professor in the Faculty of Management Sciences, Siksha O Anusandhan (Deemed to be University), Bhubaneswar, Odisha, India. He received the Ph.D. degree from SOA University, Bhubaneswar, India, and the master's degree in computer science and engineering and business administration. Dr. Dash has more than 12 years of academic experience and industry experience in India. He has received many awards for research excellence and is also a member of different institutions and societies. He has published 42 papers in peer-reviewed journals, which are indexed in Scopus, Web of Science, and listed in ABDC and more than 50 papers in national and international conferences. He has been a guest reviewer for many reputed journals. Under the guidance of Dr. Dash, four scholars have been awarded. His current research interest includes mobile banking, Internet banking, behavioral finance, waste management, and E-commerce.

Bibhuprasad Sahu received the M.Tech. degree in computer science and engineering from the National Institute of Science and Technology in 2011. He is currently working toward the Ph.D. degree with Maharaja Sriram Chandra Bhanja Deo University, Baripada, Odisha, India. His research interest is in the application of machine learning algorithms for cancer diseases. He is a member of the ISTE, IEEE, and

IEANG. He is also acting as a Reviewer for a reputed journal like *Hindawi*. He has published more than 18 international reputed journals (Scopus/ESCI).

Abhilash Pati is currently working as a Full Time Research Scholar with the Department of Computer Science and Engineering, Siksha O Anusandhan (Deemed to be University), Bhubaneswar, Odisha, India. He received the B.Tech. and M.Tech. in CSE from the Biju Patnaik University of Technology, Odisha, India. He has qualified with UGC-NET Assistant Professor only. He has more than 10 years of experience in the field of teaching in various technical and professional colleges. He has published three research articles in peer-reviewed and Scopus indexed journals and international conferences. He has been working as a Reviewer for multiple international journals and is also a member of different institutions and societies. His current research interest includes fog computing, Internet of Things, cloud computing, deep learning, and machine learning.

Sachi Nandan Mohanty received the PostDoc degree from IIT Kanpur in 2019 and the Ph.D. degree from IIT Kharagpur, India in 2015, with MHRD scholarship from the Government of India. He has edited 24 books in association with Springer and Wiley. His research areas include data mining, big data analysis, cognitive science, fuzzy decision making, brain–computer interface, cognition, and computational intelligence. He has received three Best Paper Awards during his Ph.D. at IIT Kharagpur from International Conference in Beijing, China, and the others at International Conference on Soft Computing Applications organized by IIT Roorkee in 2013. He was awarded the Best thesis award and the first prize by the Computer Society of India in 2015. He has guided six Ph.D. scholars. He has published 60 international journals of international repute and has been elected as a Fellow of the Institute of Engineers and senior member of IEEE Computer Society Hyderabad chapter. He is also the Reviewer for *Journal of Robotics and Autonomous Systems* (Elsevier), *Computational and Structural Biotechnology Journal* (Elsevier), *Artificial Intelligence Review* (Springer), and *Spatial Information Research* (Springer).

Divya Stephen received the B.Tech. degree in computer science and engineering in 2020 from Jyothi Engineering College, Thrissur, Kerala, India, where she is currently working toward the M.Tech. degree in computer science and engineering. She has published a paper in the *International Journal of Trend in Scientific Research and Development (IJTSRD)* and a few book chapters in the area of blockchain, medical image processing, and machine learning. Her project is based on medical image processing and the current area of interest includes IoT, image processing, biomedical engineering, and blockchain.

S. U. Aswathy is currently a Professor with Computer Science & Engineering Department, Jyothi Engineering College, Thrissur, Kerala, India. She received the B.Tech. degree in electronics and instrumentation engineering, the M.Tech. degree in computer and information technology, and the Ph.D. degree in medical image processing in 2005, 2009, and 2019, respectively. She has a total of 15.5 years of experience in teaching and research as a Lecturer, Assistant Professor, Associate Professor, and HOD. She is the Reviewer of many international journals. She had published more than 20 articles in various SCI and Scopus indexed journals. She had published four books in the area of machine learning and several book chapters in the same area. Dr. Aswathy's main area of research interest is in medical image processing, machine learning, nanotechnology, blockchain, and biomedical engineering.

Linkedin: https://www.linkedin.com/in/dr-aswathy-s-u-5a778618/

https://www.jecc.ac.in/departments/computer_science_engineering#tab_teaching

Priya P. Sajan is currently working as a Project Engineer with C-DAC, Thiruvananthapuram, India. She has 11 years of experience in cyber forensics and cyber security. She has handled 225+ in-house and onsite training digital forensics, 50+ training sessions on cyber security and information security for central and state government officials, premium law enforcement agencies across India, industry professionals of multiple dimensions, various universities, students, and the general public. She is a Doctoral Degree holder and is the author of 10 research papers published in various international Scopus indexed journals. Her areas of expertise also include ethical hacking, malware analysis, artificial intelligence, machine learning, IoT, and incident response. Dr. Sajan is an active trainer and an enthusiastic scholar and has obtained more than 17 certificates on multiple topics related to information security education and awareness.

Jyothi Thomas is a faculty in Computer Science and Engineering, Christ University, Bangalore, India. Her area of specialization includes data mining, machine learning, artificial intelligence, and data science. Dr. Thomas is currently working on a Government of India-funded project for devising a non-destructive method for sex identification of silkworm pupa using artificial intelligence.

S. Castro is currently working as an Assistant Professor with the Department of Information Technology at Karpagam College.

He received the B.Tech. degree in information technology from PSNA College of Engineering and Technology (approved by Anna University), Dindigul, Tamil Nadu, India, in 2010, and the M.Tech. degree in computer science and engineering from Karunya University, Coimbatore, Tamil Nadu, India. His main areas of research interest are in data mining, big data, cloud computing, and network security. He has published many papers in international and national journals. He is a Life Member of the Indian Society for Technical Education (MISTE).

P. Meena Kumari is currently working as an Associate Professor with the Department of Computer Science and Engineering, AVN Institute of Engineering and Technology, Andhra Pradesh, India. Her areas of research interest are in software engineering, database management system, artificial intelligence, and machine engineering. Totally, Dr. Kumari has 12 years of teaching experience and 2 years of research experience.

S. Muthumari is currently working as an Assistant Professor with the Department of Computer science & Information Technology, S. S. Duraisamy Nadar Mariammal College, Kovilpatti, Tamil Nadu, India. Her areas of research interests are in image processing, machine learning, and data mining.

J. Suganthi is currently working as an Associate Professor with the T. John Institute of Technology, Bengaluru, Karnataka, India. She received the bachelor's degree in special mathematics from Lady Doak College, Madurai, India, and the master's degree in engineering from RCET, Madurai, India. Having pursued a Ph.D. degree in information and communication engineering with Anna University, Dr. Suganthi's research needs extend in the domain of artificial intelligence and cognitive psychology.

About the Editors

C. Karthik is an Associate Professor of Mechatronics Engineering at Jyothi Engineering College where he teaches courses on robotics and automation, mechatronics system design, and optimization algorithms. He has 12 years university-level teaching experience in electrical and computer engineering and has a strong CV about research activities in control system design and artificial intelligence. He holds a large number of patents and has received several medals and awards due to his innovative work and research activities. He is guest editor and editorial board member for many journals and has published several international journals papers. Karthik has also made several conference presentations and worked on the development and evaluation of several interactive computing projects. His research interests include the time delay control problem, nonlinear system identification, robotics and autonomous systems, sensor design and unmanned vehicles. He was recently involved in the research of sensor design for medical autonomous systems using machine learning techniques. Karthik is a member of the Association for Computing Machinery (ACM), the ACM Special Interest Group on Computer Human Interaction (SIGCHI), a senior member of IEEE, and a member of the IEEE Robotics and Automation Society.

M. Rajalakshmi received her Ph.D. degree from Anna University, in 2020, in the area of system identification and controller tuning. She received her B.Eng. degree in electronics and instrumentation engineering from the Kamaraj College of Engineering and Technology, in 2010, and M.Tech. degree in instrumentation and control engineering from the Kalasalingam Academy of Research and Education (KARE), in 2012. She is working as an Associate Professor with Jyothi Engineering College, Thrissur, Kerala. She has published several international journal and conference papers. Her professional interests focus on machine learning, artificial intelligence, linear and nonlinear control systems, system identification, and her current projects include modeling and controlling of nonlinear process (machine learning algorithms for biomedical and robotics).

Dr. Sachi Nandan Mohanty received his post doctorate from IIT Kanpur in the year 2019 and Ph.D. from IIT Kharagpur, India in the year 2015, with a HRD scholarship from the Government of India. He has edited 14 books in association with Springer, Wiley and CRC Press. His research areas include data mining, big data analysis, cognitive science, fuzzy decision making, the brain–computer interface, and computational intelligence. Professor Mohanty received three best paper awards during his Ph.D. at IIT Kharagpur, from an International Conference at Beijing, China, and the other at International Conference on Soft Computing applications organized by IIT Rookee in the year 2013. He was awarded a best thesis award by the Computer Society of India in the year 2015. He has published 42 papers in international journals of repute and has been elected as a fellow of the Institute of Engineers, IETE, and a senior member of the IEEE Computer Society Hyderabad chapter. He is also a reviewer for Journal of Robotics and Autonomous Systems (Elsevier), Computational and Structural Biotechnology Journal (Elsevier), Artificial Intelligence Review (Springer), and Spatial Information Research (Springer).

Mr. Subrata Chowdhury is perusing M.Tech at the Sreenivasa Institute of Technology and Management Studies, Chittoor Andra Pradesh, India. He has edited 5 books in association of the CRC press and others. His research areas include data mining, big data, Machine learning, Quantum Computing, Fuzzy logic, AI, Edge Computing, Swarm Intelligence, Healthcare. He receive Awards and Nominations from the different National & International Science societies. He had published more then 50 papers in international and reputed journals. He has been the member of the IET and a member of the IEEE. He is also the reviewers for the IEEE Transactions, Elsevier's, Springers. And Academic editor for the Hindwai journals. He has been invited as a keynote speakers for many Workshops, Conferences and Seminars.